Communications
in Computer and Information Science   85

Hagen Buchwald   Albert Fleischmann
Detlef Seese   Christian Stary (Eds.)

S-BPM ONE –
Setting the Stage for
Subject-Oriented Business
Process Management

First International Workshop
Karlsruhe, Germany, October 22, 2009
Revised Selected Papers

 Springer

Volume Editors

Hagen Buchwald
Institute AIFB
Karlsruhe Institute of Technology (KIT)
Karlsruhe, Germany
E-mail: hagen.buchwald@kit.edu

Albert Fleischmann
JCOM1
Rohrbach, Germany
E-mail: albert.fleischmann@jCOM1.com

Detlef Seese
Institute AIFB
Karlsruhe Institute of Technology (KIT)
Karlsruhe, Germany
E-mail: detlef.seese@kit.edu

Christian Stary
University of Linz
Communications Engineering
Linz, Austria
E-mail: ce.win@jku.at

Library of Congress Control Number: 2010934264

CR Subject Classification (1998): D.2, H.5.3, D.3, C.2, K.6, J.1

ISSN   1865-0929
ISBN-10   3-642-15914-1 Springer Berlin Heidelberg New York
ISBN-13   978-3-642-15914-5 Springer Berlin Heidelberg New York

springer.com

© Springer-Verlag Berlin Heidelberg 2010
Printed in Germany

Typesetting: Camera-ready by author, data conversion by Scientific Publishing Services, Chennai, India
Printed on acid-free paper   06/3180   5 4 3 2 1 0

Foreword

This volume contains a selection of papers from the First Workshop on Subject-Oriented Business Process Management (S-BPM ONE). Establishing a multi- and cross-disciplinary interchange of underyling and applied concepts, successful application studies, and innovative development ideas, the workshop emphasized the proactive realization of role- or actor-oriented modeling on the basis of exchanging messages when accomplishing tasks.

The workshop was organized as a forum for the discussion of foundations, achievements, reflections, and further developments. In this way, its contributions not only addressed the current state of the art, but also the various lines of research and development, either running or planned. The state of the art is reflected in terms of concepts, modeling language, and tool features on the one hand. On the other hand, it is reflected through the discussion of industrial case studies. These indicate the current practice when implementing the subject-oriented BPM paradigm in industrial settings. By challenging conceptual foundations they also allow us to define a common ground for future developments in research and practice.

The S-BPM ONE contributions focus on challenges arising from the evolution of service-oriented architectures and the need for more flexible business organizations. The latter require coherent and adaptive representation and processing techniques for business process modeling and execution. Corresponding technologies have to be grounded in theories of computer science, in order to provide an adequate infrastructure for thorough BPM including technology-enhanced change management.

The contributions do not only review the basic concepts and business-relevant applications of subject-oriented BPM, they also provide substantial evidence of the third wave in BPM. The findings have been grouped according to envisioned S-BPM implementations including the required paradigmatic shift, the capabilities of S-BPM to establish semantic enterprises, and the next steps that need to be addressed in S-BPM research and application:

- Part I (Visionary Engagements) indicates the need for a paradigmatic shift towards S-BPM, and provides practical and conceptual evidence, looking at business operations and applying systems thinking. The various inputs do not only take into account current developments, such as the diffusion of Service-Oriented Architectures into the Internet for S-BPM, but also the demand for S-BPM education programs and training environments.
- Part II (Essential Capabilities) gives an overview of the state of the art in S-BPM, addressing the shift to semantic support technologies. Besides the fundamental concepts and inherent capabilities, industry-relevant implementations of subject-oriented task scenarios are detailed.

- In part III (Penetration Perspectives) further developments in S-BPM are discussed. They range from organization design to technology improvements for networked organizations. The most urgent issues to advance S-BPM could be identified in different formats at the workshop and have become part of a multi-dimensional S-BPM road map.

In part I Lutz Heuser's contribution sets the stage in terms of enterprise resource planing for agile organizations. His experiences in research and development demonstrate the crucial role of innovative services in BPM, in order to keep up with organizational and technological alignments of competitive enterprises. A key enabler is semantic processing which is also reflected by Hagen Buchwald revisiting implemented business processes rather than focussing on envisioned ones when introducing S-BPM. Epistemological analyses also enables, as demonstrated by Christian Fichtenbauer, that there still exists a variety of hindrances to implementing business processes in socio-technical systems. According to his findings, inherent system characteristics, such as operationally closed action cylces, form barriers for straightforward process specifications and implementations.

Education and skill development have to be considered crucial success factors for S-BPM. Werner Schmidt and Christian Stary propose to streamline education and training efforts ensuring quality, continuity, and transparency of development. Robert Singer and Erwin Zinser provide insights in current S-BPM teaching and training in the academic setting of applied sciences. Their data reveal that effective and sustainable embodiment of subject-oriented modeling and management into existing curricula requires substantial programmatic effort. However, its benefits can be shown for industrial applications, as Erwin Aitenbichler and Stephan Borgert demonstrate by successfully increasing business intelligence when processing subject-oriented representations. The case is detailed in the concluding section of part I.

Part II is opened by Albert Fleischmann providing a structured review of S-BPM developments towards semantic modeling and processing. The language and essential features of the tool for executing subject representations are detailed, before two case studies are presented. Anton Kramm shows in his case study, life cycle requirements when generating portals and complex event processing applications. For the latter executable processes form the backbone of organizational behavior. Moreover, S-BPM in combination with other technologies increases the agility of software development and implementation. Hereby the orchestration of IT services and their adaptability play a crucial role, besides work rules, patterns of behavior, and events triggering process execution.

The second case study is provided by Gabriele Konjack. It illustrates an application of S-BPM and its technology infrastructure, the JCOM1! Suite, for subject-oriented order control in financial services. The field work enlightens not only substantial modeling tasks, but also the management perspective. The case supplements the content-driven perspective detailed in the first case study with project management tasks required for implementing change. Management has still to be studied and explored with respect to support instruments.

Open issues are also part of Hagen Buchwald's roadmap design in the initial paper of part III, as well as the inputs provided by all participants in the lively discussion at the First S-BPM World Café. The roadmap captures the technological, community, and methodological building blocks of S-BPM. For each category a fundamental set of activities considered to be crucial for development and research is provided. Some of them were directly addressed at the World Café. Important issues are standard setting, cultural embodiment, and education, besides working business plans for S-BPM. Performing this collective activity the economic, educational, social, and content perspective on S-BPM could be revisited and aligned with the inputs previously provided. The appreciated contributions to the World Café were enriched by an oral epilogue given by Detlef Seese.

We would like to thank all the knowledge activists and the organizers for doing their best in realizing this milestone, setting the stage for fundamental and applied research in subject-oriented BPM. We are convinced that the results will not only provide in-depth understanding of existing concepts and applications, but will also accelerate human- and business-centred implementations of the third wave in BPM, empowering stakeholders and networked organizations.

July 2010

Hagen Buchwald
Albert Fleischmann
Detlef Seese
Christian Stary

Table of Contents

Part I

Visionary Engagements

The Relevance of Management of Business Processes and Orchestration

Lutz Heuser

SAP Corporate Research
Dietmar-Hopp-Allee 16, 69190 Walldorf, Germany
Lutz.Heuser@SAP.com

Abstract. Developers committed to subject-orientation are able to increase Business Process Management's (BPM's) efficiency and effectiveness. First, the modeling of processes becomes much more efficient due to the coherent creation of main processes, once responsibilities are clarified. Secondly, remaining issues are more transparent for project managers and controlling, as the states of projects become visible and traceable. Thirdly, there are fewer manual actions required for modeling and management, due to focused message passing and communication. Finally, the integrated, computer-based workflow implementing such type of specifications guarantees a more efficient and faster process flow.

Keywords: Business Process Management (BPM), subject-oriented modeling, Service-Oriented Architecture (SOA), orchestration, collaboration, Cloud Computing, Workflow Management.

1 Introduction

Starting in the early days the experiences of (SAP-)consultants indicate that both, translating business process models in such a way stakeholders understand them, and the (software) system implements them, is crucial. Why not empowering users to explain what they need in terms of system support? This idea is today more or less a discipline of its own – end-user development.

I personally believe that this is the way to go. The software industry, also SAP, is facing a critical time since enterprise applications are becoming more and more of a commodity. We anticipate not any longer large implementation projects, but rather the increased usage of SaaS (Software-as-a-Service) and Cloud Computing – software industry has to respond to that. S-BPM is one of the ways to successfully address this ongoing change.

2 European ICT Strategy 2020

The European ICT Strategy 2020 is very important for SAP Research and drives the activities through research projects. The megatrends identified by the European Commission, as well as by SAP, around the future Internet are Cloud Computing,

H. Buchwald et al. (Eds.): S-BPM ONE, CCIS 85, pp. 3–12, 2010.
© Springer-Verlag Berlin Heidelberg 2010

Internet of Things, and Internet of Services. S-BPM fits perfectly to the latter which is about how to compose and combine services in order to support cross-domain, cross-enterprise business processes.

From the societal perspective it is important to attain a sustainable economic recovery. What has to be kept in mind is that ICT has contributed to the crisis by creating the environment that allowed this crisis to occur. New business models, like SaaS (Software-as-a-Service) and Cloud Computing, will emerge. These will not only be fueled by ICT, but will also change the ICT sector in itself. The software industry will cease to be a capital investment and will instead become an operational expenditure. And this will change the business models which will in turn represent a great opportunity for Germany because of its long tradition in combining computer science with business. Additionally, ICT is aggressively moving towards achieving new global capabilities with regards to embedded, connected, ambient systems. We also have to understand that the cloud – the fact that everything is going to be connected - will provide us with much more insight into the real world. So data will be available and this data has to be used in the context of the business process. We also believe that this will lead to a fundamental shift in the way business processes are set up in the future. The idea and belief is that in the future enterprises will become so called "digital enterprises". This means they will determine their business processes in accordance with the fact that they have IT embedded. So rather than using IT to optimize business processes, we are going to develop business processes with this concept of embedded IT in mind. This is exactly what S-BPM does: It addresses embodied IT in organizational settings.

Last but not least, from a European perspective we must clearly understand that we are not on our own. First of all, we should realize that the European Union constitutes the largest ICT market in the world. We are larger than Asia, we are larger than Russia and we are larger than the US. But we are importing ICT from countries outside the European Union. This means European companies have a great opportunity just within their own market, but yet have to understand that this is not an enclosed environment. We are not living in medieval castles any longer, it's an open innovative world and we have to collaborate. Thus, it is very important for us to take into account the evolving new champions – often called the G20. - I have already listed a couple of these champions above who are already present and who have a clear understanding of the fact that they are addressing Europe as a market, addressing ICT in the European market. So, we are looking for ways to collaborate more strongly with these partners in the future. Within the EU, through Estac we already propose sustainable collaboration by means of joint calls and hope that this will further mature. Joint calls means that the EU together with international partners will provide as members joint funds for research to enable and promote collaboration among these member countries themselves as well as between the member countries and the countries listed above (including USA, Australia, Brazil, China, Russia, India).

Additionally, we will see an increase in the establishment of joint think tanks among these countries and Europe to support a more extensive exchange of these ideas which will promote a much tighter integration.

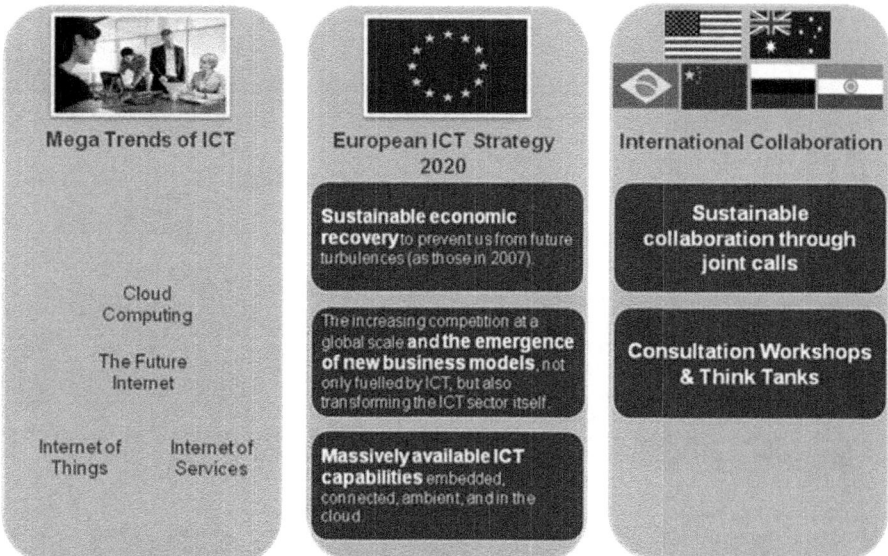

Fig. 1. Strategy development

3   Future Internet

Now let's have a look at the Future Internet and what it will mean for us. The Figure below was developed by Estac about two years ago and SAP Research is constantly using it to promote its ideas.

First of all, if you look at the lower level you see what people like to call the network layer. Here, we need to be aware of the fact that we are in the advanced phases of a conversion process. Some people think this has already been completed, but however, if you take a more detailed look at embedded systems, you will see that a lot of proprietary network infrastructure still applies and needs to be overcome. At the end of the day we need to have IPv6 (Internet Protocol version 6, or whatever the future network protocol will be) everywhere. Ultimately, a harmonized infrastructure will evolve to support anybody, anytime, anywhere with anything. We can see its impact everywhere. There are billions being invested to this effect, so this goal will not disappear, it will be achieved. This fact is enforced by current topics behind buzz words like electro mobility or eEnergy. SmartMeters will have an IP address, cars will have several and there is no escaping this. That's why we also have to clearly understand that IPv4 (Internet Protocol version 4) with its inherent limitations (such as the limited number of remaining available IP addresses) will become obsolete at some time in the foreseeable future. Networks will no longer function in this context. We need to have direct access to all of these devices (end-to-end principle) and thus we need to have unique addresses for each of them. So the Internet of Things will largely drive internet convergence and the introduction of IPv6 as a whole.

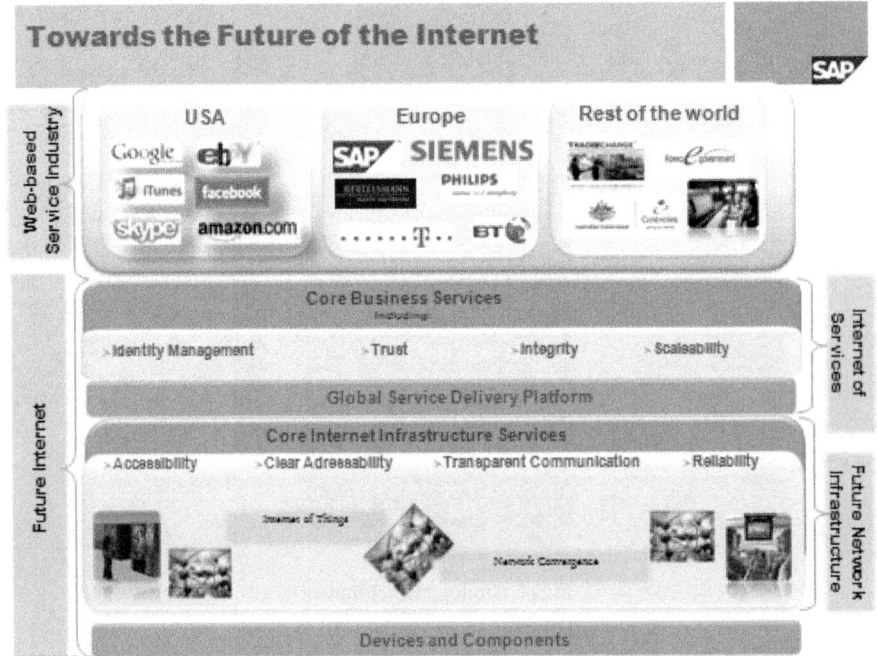

Fig. 2. Internet developments

In addition to this you can see a number of services we need to have in order to make things happen. And subsequently we reach the topic we are discussing today – the main services required to do business process modeling and then ultimately, business process management. When we talk about business process management we are speaking about a cross-enterprise and not an intra-enterprise issue. So, we need to have a clear understanding of whom we are collaborating with, of what the level of trust is that we can achieve among them and of how the integrity of the whole process and, in particular, the data obviously coming into play need to be on internet scale.

So, many issues need to be addressed and this is what we refer to as the Internet of Services.

4   The Role of Business Process Management in the Future Internet

First we should consider the emerging market of on-demand BPM solutions as a result of Cloud Computing and the Internet of Services. In the past 20 years (business) process discussions were largely dominated by a notion of workflow management. Today we have to clarify whether we are really talking about BPM or more about workflow management being referred to as BPM. For me, BPM goes far beyond workflow management and in turn, S-BPM represents true BPM since S-BPM is about end-user (or business user) development and empowerment. This is the profound difference between WFM (Work Flow Management) and BPM. Most of today's so-called BPM-tools are addressing technical issues rather than business users and should therefore more appropriately be called workflow tools.

Gartner estimates that the market for BPM systems and BPM-enabling technology will reach 3.6 Bill US$ in 2011. It has the highest growth in the outsourcing industry with a Compound Annual Growth Rate of 10% and Forrester claims that BPM technologies will be the most frequently implemented and updated technology category in 2009. For the BPM outsourcing market Gartner estimates market size to reach 172 Bill US$ in 2009 with the highest growth in the BP outsourcing industry with a Compound Annual Growth Rate of 9.1%.

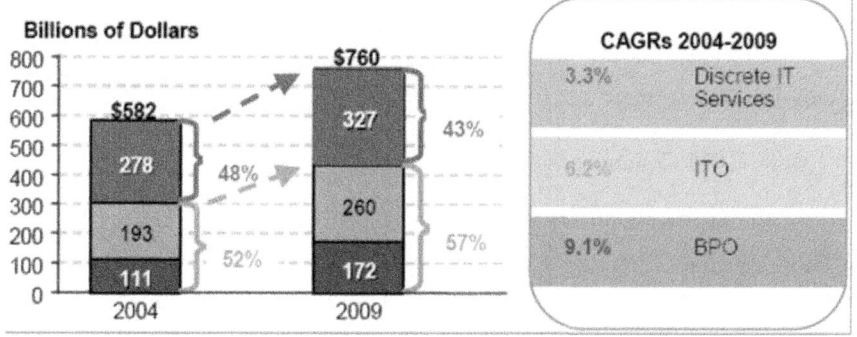

Fig. 3. Market developments

5   Research Vision

In SAP research I see four main areas of work: bring BPM into the cloud, make BPM lightweight, make BPM collaborative and make BPM a commodity.

5.1   Bring BPM into the Cloud

Our vision of why this should happen is based on the low entry costs and the quick ROI which drives Cloud Computing.

- This accounts for a shifting of buying power in the customer base by addressing LOBs.
- It effectively leverages SAP Research's vision of Internet of Services by
  - the ability to invoke registered and brokered third party services into cloud-based processes
  - the capability to wrap cloud-based processes as services and register them
  - enabling service providers to utilize service charging facilities.
- This approach allows easy process integration along supply chains due to a common infrastructure.
- It supports connections to enterprise systems behind firewalls through secure data connectors.
- It permits resolution of on-premise events through cloud-based scalable event hubs.

5.2 Make BPM Lightweight

The time of very massive implementation measures and lengthy on-going and very costly software projects has passed and end-user empowerment will emerge. This is especially true because of the fact that client technologies need to be end-user environments (built to be easily and directly used by general business users). Typical examples for these technologies are browsers and applications such as Facebook, existing end-user applications like MS Outlook and future applications such as Google Wave which require neither installation nor cumbersome adaptation. Similar to these BPM will develop to the point at which it will require no programming and implementation efforts. It will consist of drag and drop operations and drawing connections. The infrastructure will mainly consist of distributed and incorporated execution environments without centralized servers. S-BPM has to some degree paved the way for this development.

5.3 Make BPM Collaborative

Collaboration is becoming a key feature.

- "Processpedia"
    - Processes are described by the people participating in a process. Each party describes their responsibilities within the process and coordinates these with their direct communication partners.
- BPM needs to evolve from within an organization and not from outside
    - Currently, there is a common use of external highly paid consultants which results in internal acceptance problems, etc.
    - There is a need for effective end-user collaboration mechanisms:
        - specification methods which can generally be easily understood without extensive trainings
        - corresponding tools which can be directly used
- Collaboration along the entire BPM lifecycle:
    - Design: high level view of process, UI design and Web Service binding
    - Deployment: on the fly or one click operation through cloud-based infrastructure, smooth inherent transition between design and deployment
    - Execution: within end-user environments, collaborative
    - Analysis: share, collaborate, discuss, collaborative brainstorming to reiterate and redesign the process
    - ➔ All in one spot, as collective intelligence.

5.4 Make BPM a Commodity

- Simplify BPM
- Find the 10 percent of technology that accounts for 90 percent of all use cases
- Commoditize languages:
    - Make use of BPMN 2.0 (explicitly or implicitly) as a standard with dedicated execution semantics.

- Commoditize execution environment:
  - Provide app as a gadget to be executed in any end-user environment, similar to the YouTube-embedded URL that allows you to invoke a YouTube video within any website or as a gadget within all sorts of applications.

5.5 BPM&SI – The Big Picture

BPM for end-users has to be considered the core of all organizational development.

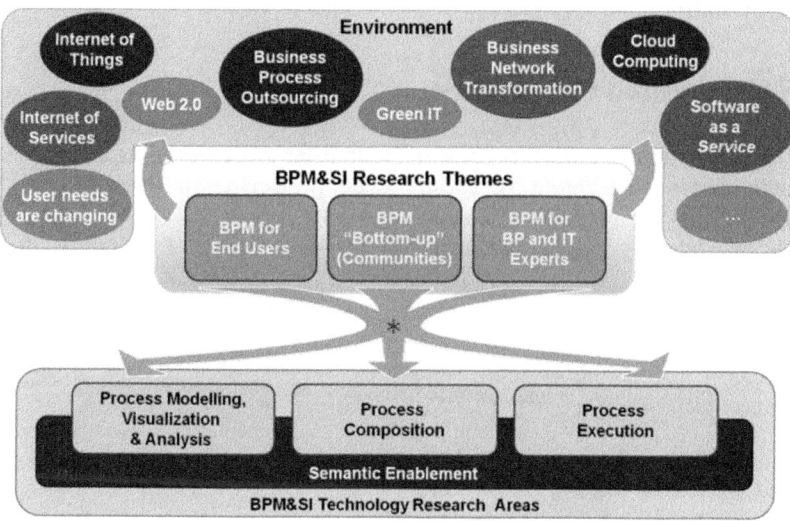

Fig. 4. Research Areas

5.6 Examples

Integrated cloud-based collaborative end-user tool BPM tool suite → Enterprise 2.0 made a reality:

- "Gravity" for high-level collaborative modeling in BPMN
- "Rooftop4BPM" (aka "Skyline") for collaborative application development
- "Slipstream" for real-time end user process analysis
- "Yowie" for smooth process execution without tool breaks

From a theoretical model to a live environment

The major Changes in Business Process Management:

- From process analysis and process design to execution-oriented process management
- Factors - process scenarios - challenges - resolution methods
- Portals, BPM, and SOA are growing together

The technology implemented:

- Innovative Business Process Management with a subject-oriented, methodical approach (S-BPM) and LIVE PROGRAMMING in the operating (business) department
- Modeling – validation and immediate execution portals

The Solution:

SAP Research is analyzing S-BPM in the context of transfer projects processes: Application, staffing, and approval of complex research projects.

6   SAP Research Core Process

SAP Research's most critical process is the transfer-project process at the end of the research process (see Figure 5). These process projects are initiated for transferring research results into products or customer projects.

The following figure gives a rough overview of the goals of the process for transferring projects.

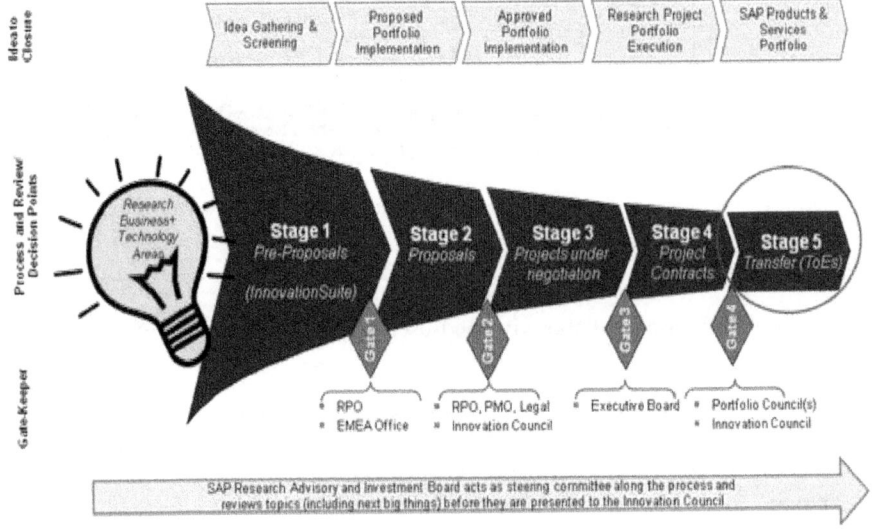

Fig. 5. Goals of project transfer process

Based on this process outline the communication channels are identified and described.

This so-called TOE (Terms of Engagement) process is used between SAP Research and Product Development when applying for internal research projects. Initiators (Transfer Project Lead) can be employees on either side. The project is then coordinated between the organizational units involved both in terms of its content

(SAP Target Group Content) and its budget (SAP Target Group Budget via Business Development). If these coordination efforts are successful, then the project resources are arranged in collaboration with the heads of the individual research laboratories (Center Management).

The communication diagram is shown in the figure below. It was modeled by subject matter experts, experts from the project management office and the customers of the process within SAP.

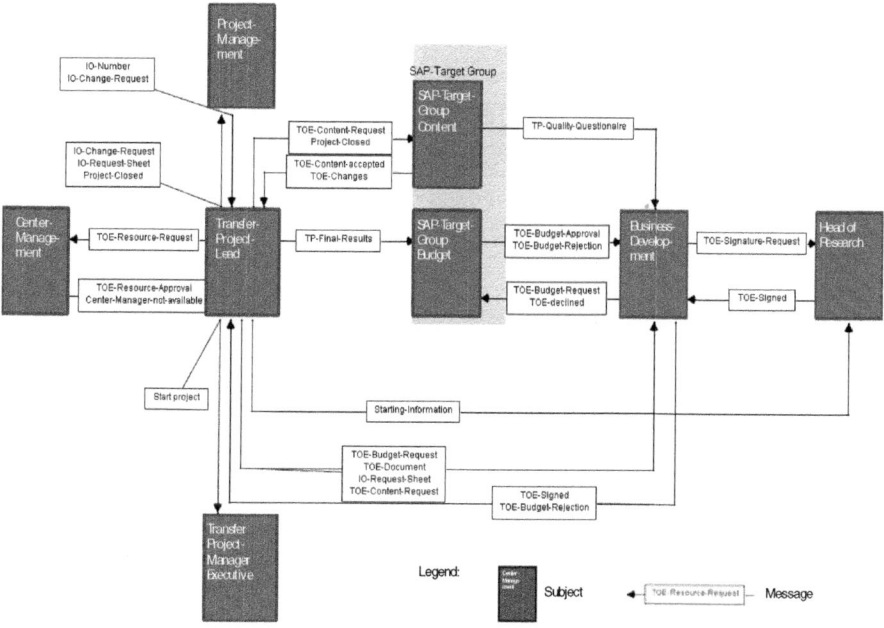

Fig. 6. Transfer project communication diagram

For instance, the Transfer Project Lead, as the initiator records the TOE document (TOE document ready) after the discussion of the project, sends the starting information to the research manager (the Head of Research), and also sends a TOE-Content-Request to the subject "Business Development" for informational purposes. The beginning of the process for the Head of Research is the receipt of the starting information from the Transfer Project Lead. The approval request is then received from Business Development, etc. One example for a refinement in the behavior of the Transfer Project Lead is a service in the send state, namely "Inform Head of Research." This service has the effect of automatically extracting the starting information to be sent from the TOE document and inserting it into the message to the Head of Research.

Based on a communication-based process model, executable workflows could be derived to a great extent. This executable process model was embedded into the organization and IT-Infrastructure of SAP Research.

7 Conclusive Findings

Due to subject-orientation the modeling of processes is much more efficient. This is emphasized by the fact that the main process was created in just one day with the respective responsibilities. The open project requests are more transparent for project managers and controlling. The overview of the states of all projects is better. There are fewer manual actions since text documents are no longer necessary due to the fact that emails are generated instead. The integrated, computer-based workflow guarantees a more efficient and faster process flow.

The Power of 'As-Is' Processes

Hagen Buchwald

Institute AIFB, Karlsruhe Institute of Technology (KIT)
Kaiserstraße 1, 76128 Karlsruhe, Germany
hagen.buchwald@kit.edu

Abstract. This article answers the question of why there is a need for S-BPM. It motivates the economical, psychological, technological and evolutionary drivers of Subject-oriented Business Process Management (S-BPM) – and the respective perspective on S-BPM. Besides improving the productivity of work based on processes as perceived by stakeholders, S-BPM implements choreography to synchronize a self-organized pattern of work processes.

Keywords: stakeholder perception, 'as-is' business process, division of work, dice game, Küber-Ross curve, Moore's new law

1 Why S-BPM?

Why do we think that 'as-is' (i.e. currently implemented) business processes are more powerful than so-called to-be or industry-best-practice processes? And what is the role of this new process modeling paradigm called S-BPM (Subject-oriented Business Process Management)? Four observations help to answer this question: an economical one, a psychological one, a technological one and a historical one.

2 The Economical Rationale

Observation 1. BPM projects often have the wrong goal!

In 2002/03 a remarkable BPM project was accomplished at a bank called Bancafé in Bogota, Columbia. The remarkable thing about this project was that the CEO of this bank, Mr. Pedro Nel Ospina, followed the theory of constraints established by Dr. Eliyahu M. Goldratt [2].

Bancafe had a cost problem. But the CEO did not react as so many economic leaders tend to react. He did not start a cost cutting program. Instead, he initiated a BPM project with the goal to reduce cycle times drastically.

H. Buchwald et al. (Eds.): S-BPM ONE, CCIS 85, pp. 13–23, 2010.

May 2003: Our desparate Consultant jets to Colombia. He has heard of a bank which needed a miracle.

„A bank is a decision making body."

„Everybody was busy,
but nobody was doing business."

„We had 250 branches – und 250 different credit processes."

„People were solving internal problems –
and not the problems of our customers."

„We had tons of papers piled up in our offices."

Pedro Nel Ospina, CEO Bancafé
Bogotá, May 2003

„We had too high costs – and still department heads were
demanding even more staff in order to be able to deliver faster."

„Information was a privilege."

„Nobody was able to articulate the whole process."

Fig. 1. Typical starting point

**„A headache became business!". How to turn a
hazardous challenge into the chance of a lifetime?**

Fig. 2. Looking for anchors

**The core concept: Improve cycle times! End-to-end
business process measuring from the branch ...**

Fig. 3. Targeting

**... down to the headquarters in Bogotá: Everybody
knows, who is responsible for each step and decision.**

Fig. 4. Transparent implementation

The effect: Cycle Times, i.e. for a credit card application & delivery process, were
reduced by more than 75% - and simultaneously costs went down by more than 50%!

In the result, cycle times were improved by 70%.
And as a side effect, costs were cut by more than 50%.

Fig. 5. Results

„The division of work shall improve the productivity of
work more than any other means."Adam Smith (1776)

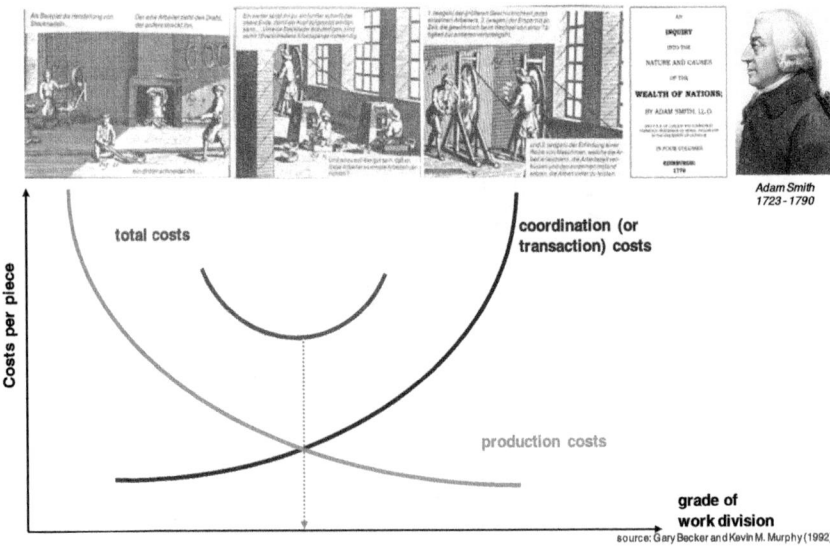

Fig. 6. Dividing work

Hypothesis 1. BPM has the mission to improve the productivity of work by enabling employees to divide their work without raising the coordination costs.

Costs are not a root cause, but a symptom! To work on costs means to work on symptoms – this is neither effective nor efficient.

Production costs are a function of three variables in the enterprise: throughput, backlogs and cycle time. Throughput, backlogs and cycle times themselves are key performance indicators of how efficient the division of work is coordinated.

By keeping the throughput constant and simultaneously reducing backlogs and cycle time, the CEO of Bancafe succeeded in cutting down costs in a sustainable manner.

But how did the CEO of Bancafe decrease backlogs and cycle times?

The answer is: By improving the coordination of the division of work.

What does this mean? To understand this, we have to look back about 250 years. In 1776 Adam Smith stated in his famous work "An Inquiry into the Nature and Causes of the Wealth of Nations": "The division of work shall improve the productivity of work more than any other means!" This is true for simple products – like pins. Adam Smith described a productivity boost by the division of work in a pin manufacturing company from 200 pins per day up to 48.000 pins per day – produced by the same number of workers!

In the early twentieth century Henry Ford applied this principle of work division on a more complex product – a car, the famous Ford T Model. To his surprise he observed that he had to heavily coordinate the divided work stations in order to get the benefit of this division of work. This coordination was so expensive that it threatened to consume the benefits of the division of work. His invention: He adopted the conveyor belt principle he had seen at Chicago's slaughterhouse to his car factory – with tremendous success. That was the birth of what we call Taylorism.

What does the conveyor belt do in order to decrease the coordination costs?

Eliyahu Goldratt [2] described the effect of a conveyor belt in his book "The Goal" by means of a simple game: the dice game, which shows in an impressive way the effect of work division in a supply chain as a sequence of interdependent tasks underlying stochastic variability (simulated by dice).

Bancafe's BPM system did for Pedro Nel Ospina what the conveyor Belt did for Henry Ford: It reduced the variance of cycle times at each working station and therefore reduced backlogs which in turn reduced the cycle time of the end-to-end process.

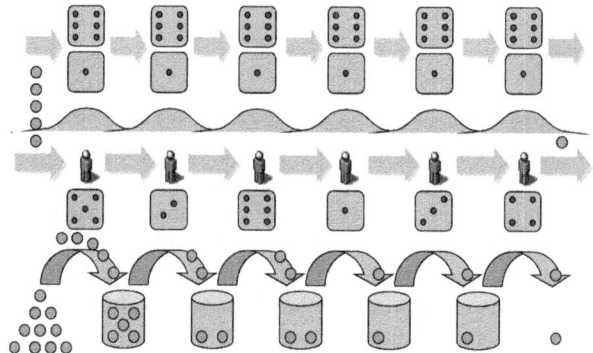

Fig. 7. Propagating reduction of cycle times

One Bancafe employee told the author of this article what the key benefit of the new system was for him: "I know who is responsible!" That is the simple truth behind the dice game: Each employee has a clear responsibility which includes the goal to deliver in time, so that oscillation of the value chain is eliminated.

3   The Psychological Rationale

Observation 2. BPM projects are prone to fail – even more so than classic IT projects (which fail far too often)!

The main reason why BPM projects often fail is the resistance they create by raising the so-called Kübler-Ross curve [3] which describes the psychological effect of how human beings react in the face of loss. We could observe the Kübler-Ross curve in a number of BPM projects we performed in the classical style, namely trying to introduce a so-called industry-best-practice process.

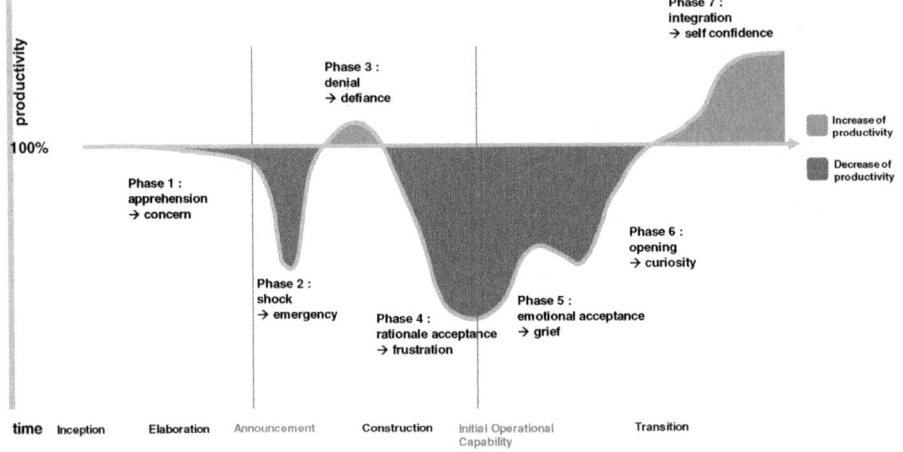

source: Dr. Kübler-Ross, 1969

Fig. 8. Increase / Decrease of productivity – classic approach

The above mentioned project in Bogota did not introduce a 'to-be' process. Instead the CEO insisted on implementing the as-is processes and only to change them based on measured data and together with the people working in the process. The surprising effect: The project actually did only evoke minor resistance in the beginning.

After the first presentation of prototypes to the employees, the lurking resistance changed into open support for the project. The reason could be comprised in one sentence. When asked, why she supports this BPM project so strongly, an employee gave the following answer to the author of this article: "But it is our process! You understand: Our process!" And this statement shows what kind of loss it is that employees fear: It is the loss of their habits! - Habits of how they did their work up to now; habits which made them successful; habits which made them valuable for the company.

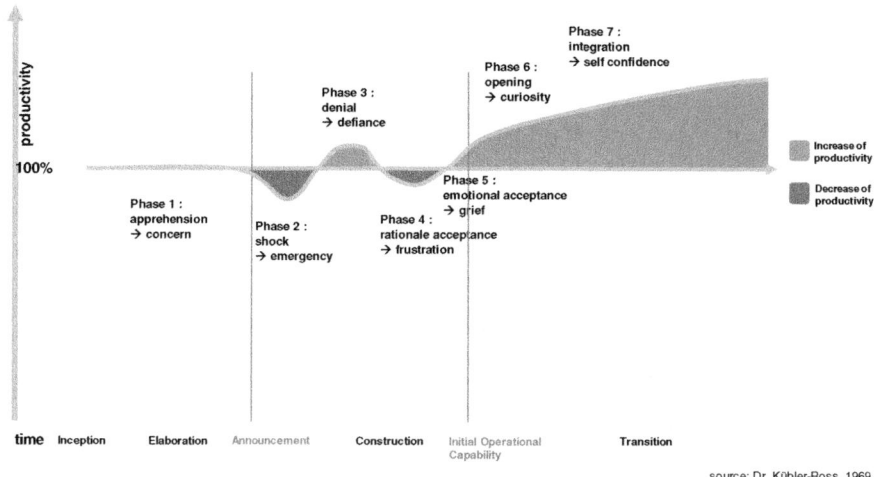

Fig. 9. Increase / Decrease of productivity – neo-classic approach

They are going to lose most of them if an externally envisioned process is introduced, since it tells everybody what he or she has to do when, how, where and why.

Hypothesis 2. BPM projects focusing on as-is processes have much better chances to achieve their goals.

Trying to introduce so-called industry-best-practice business processes raises the Küber-Ross [1] curve, which means that resistance against the BPM project increases. This resistance leads to a severe loss of business performance – an unconscious reaction of the employees affected by this project.

This loss of business performance can't be ignored by responsible managers. Therefore they tend to give the project a sweet death: Let the project die silently before it gives us a sudden death.

4 The Technological Rationale

Observation 3. Moore's law has changed. The classic version of Moore's law said that computing power would double every 18 months. The new version of Moore's law claims that the number of cores per chip currently doubles every 24 months and soon will double every 18 months!

Doubling the number of cores is not equivalent to doubling the computing power of single core systems. Moore's classic law doubled the performance of our applications every 18 months without altering a single line of code by simply installing these applications on the next generation hardware. This software engineers' paradise has come to an end. In order to use the parallel computing power of multi-core hardware, software has to be inherently parallel itself!

Mass Production starting from	2004	2006	2008	2010	2012	2014	2016	2018
Technology (nm)	90	65	45	32	22	16	11	8
Number of Cores per Chip	2	4	8	16	32	64	128	256

already superannuated!

**New Form of Moore´s Law:
Number of Cores per Chip
doubles every 24 months!**

- ⇨ hundreds of Cores per Chip in near future !
- ⇨ affordable for everyone !
- ⇨ applications itself must be inherent parallel !

KIT IPD is at the top
of this revolution !
⇨ XJava, Parallel Design Patterns
source: Prof. Dr. Walter Tichy, KIT, IPD

Fig. 10. A new law for chip development

The first generation of BPM systems belongs to the so-called ARIS class – you model business processes, but you can't execute this model.

Second generation BPM systems model and execute business processes – and thereby measure key performance indicators like throughput, backlogs and cycle times. The above mentioned Bancafe project was based on such a second generation BPM platform, which we call internally the BizAgi class (short for business agility).

But these second generation BPM platforms have one thing in common: They conform to an orchestration paradigm. And just like in an orchestra there can only be one conductor. This conductor becomes the bottleneck when Moore's classic law is substituted by Moore's new law!

Hypothesis 3. The third generation of BPM systems will be inherently parallel in order to cope with Moore's new law. These BPM systems therefore will conform to the choreographic paradigm, rather than the orchestral paradigm which dominates BPM systems today.

5   The Evolutionary Rationale

Observation 4. Natural language is our means to coordinate the division of work. This capability – given to human beings by their natural language – is the distinguishing evolutionary advantage human beings possess.

Are programming languages like natural languages?

In 1970, Wirth and Dijkstra introduced the structured programming paradigm, which concentrated on the functions. Compared with the way we speak one could say: The structured programming paradigm concentrated on the predicates of our natural language. The so-called top down approach helped to reduce system complexity by a stepwise, systematic (structured) refinement process over different levels of abstraction until the problem was concrete enough to be solved in a single function.

This structured programming paradigm worked fine in the first part of the life cycle of software systems: analysis, design, implementation and even test. Problems occurred when these systems were successful. Success raises the demand of the system users for more functionality. Introducing this functionality into the top down structured design turned out to be a rather complex task, which could not be solved by the structured programming paradigm. Modifying top-down design systems was error prone – each new feature increased the functionality and decreased the quality of the system. The effect: Maintenance costs exploded – the system became a costly legacy.

In 1990, the object-oriented paradigm had its breakthrough. Bertrand Meyer made the observation that during the lifecycle of a system it is not the functions which are stable. It is the data structures which do not change even if new functions have to be introduced. Consequently, he focused on the objects and made the functions part of the objects. Using the metaphor of comparing a programming language with natural human language you can say: He did not focus on the predicate, but rather on the object of a common SPO-formed sentence. SPO-formed means: subject – predicate – object.

This is the natural way we speak. And we use our language to deal with the complexity of work division! Language is the differentiating tool of human beings as opposed to animals, as it allows us to master the complex dynamic phenomena resulting from the division of work and shown above in the dice game.

However, the object-oriented paradigm only allows us to talk in mutilated sentences: object – predicate.

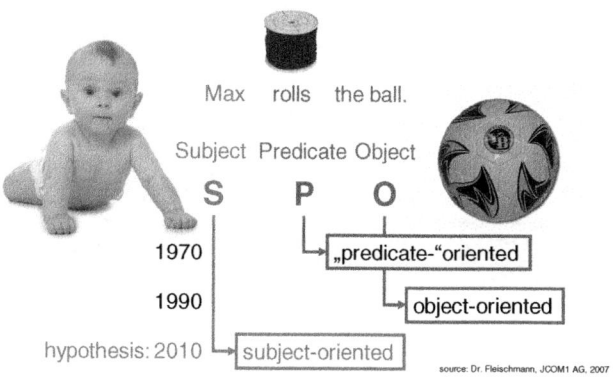

Fig. 11. Subject – Predicate - Object

1970 – 1990 – 2010 – it's time for the next evolution level of programming languages. And this leads us to hypothesis 4:

Hypothesis 4. A programming language, that naturally helps human beings to improve their productivity of work by coordinating the division of work, must offer all three parts of a natural language sentence: subject – predicate – object. S-BPM – the BPM of the third generation – will be such a natural programming language.

Remark: This hypothesis is flanked by another hypothesis:

Hypothesis 5. Changes of paradigms in a scientific field correlate with changes of their "popes" in that field (alternation of generation). And one generation in IT science is 20 years.

Hypothesis 4 directly leads to the question: What exactly is S-BPM?

There will be a first attempt to answer this question in the article about the roadmap to S-BPM [1].

The other question is: Will the subject-oriented paradigm replace the object-oriented and "predicate"-oriented paradigm?

The answer will be given with our final hypothesis:

Hypothesis 6. The evolution of programming languages reaches its third age. All three paradigms will coexist – but each of them on that particular layer of multi-tiered systems, promoting their individual strengths.

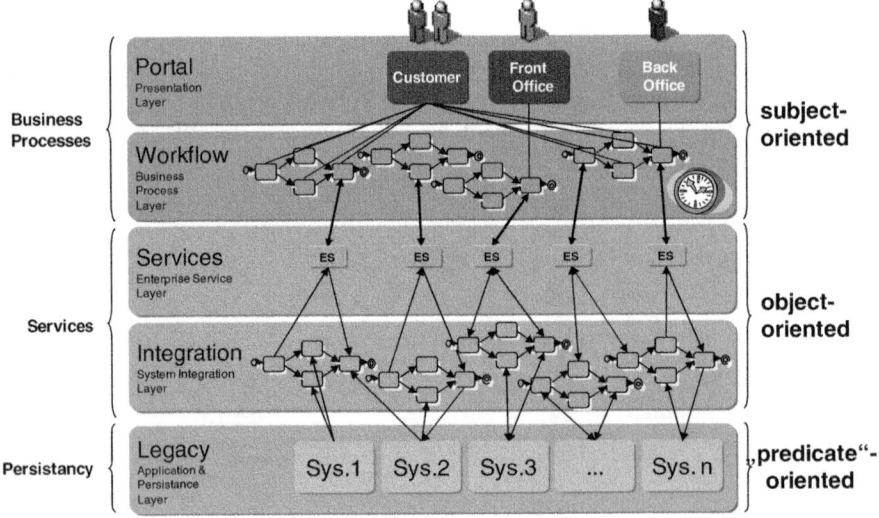

Fig. 12. Third age scenario

6   Conclusive Summary

This article states four observations of why there is a need for S-BPM as a new programming paradigm:

1. BPM projects often have the wrong goal!
2. BPM projects are prone to fail – even more than classical IT projects!
3. Moore's law has changed. The classic version of Moore's law said that computing power would double every 18 months. The new version of Moore's law claims that the number of cores per chip will double every 18 months!

4. Natural language is our means to coordinate the division of work. This capability – given to human beings by their natural language – is the distinguishing evolutionary advantage human beings possess.

This article also states four rationales, in the form of hypotheses, about what these observations mean to S-BPM:

1. The economical rationale: BPM has the mission to improve the productivity of work by enabling employees to divide their work without raising the coordination costs.
2. The psychological rationale: BPM projects focusing on as-is processes have much better chances to be achieve their goals.
3. The technological rationale: The third generation of BPM systems will be inherently parallel in order to cope with Moore's new law. These BPM systems therefore will conform to the choreographic paradigm, not the orchestral paradigm which dominates BPM systems today.
4. The evolutionary rationale: A programming language, that naturally helps human beings to improve their productivity of work by coordinating the division of work, must offer all three parts of a natural language sentence: subject – predicate – object. S-BPM – the BPM of the third generation – will be such a natural programming language.

Acknowledgements. Special thanks go to Dr. Albert Fleischmann for stimulating discussions on the very first ideas of this subject during several meetings starting in 2003 and to Professor Andreas Oberweis and Professor Detlef Seese of AIFB, KIT for their deep questioning in trying to explore the real nature of S-BPM.

References

1. Buchwald, H.: Potential Building Blocks of S-BPM. In: Buchwald, H., et al. (eds.) S-BPM ONE. CCIS, vol. 85, pp. 123–135. Springer, Heidelberg (2010)
2. Goldratt, E.M., Cox, J.: The Goal. North River Press, New York (1984)
3. Kübler-Ross, E.: On Death and Dying. Scribner Classics, New York (1993)

The Method behind Subject Orientation – The Missing Link between Individuals and Machines in Regard to Truth

Christian Fichtenbauer

Dr. Fichtenbauer EDV- & Prozessorganisation GmbH,
Schulgasse 44/6, 3051 St. Christophen, Austria
Christian.Fichtenbauer@cf-consult.at

Abstract. Given the universe of discourse of S-BPM, namely socio-technical systems, the perceived reality of stakeholders involved in work processes play a crucial role. Once their (daily) experience in those processes can be captured adequately, the adaptation of processes to a changing environment or to novel ideas for accomplishing tasks can be facilitated. By putting actors and their communication behavior to the center of interest subject orientation is a promising candidate for the symbiotic organization of work in socio-technical systems. The immediate execution of specifed models has to be considered another key enabler to that respect, as it meets the demand for ‚What You Specify Is What You Get' in a straightforward way.

Keywords: socio-technical system design, stakeholder perception, observer dilemma, self organization of work, subject orientation

1 About Truth

Let us start with a citation of Ernst von Glasersfeld ([8], p.1): "What is radical constructivism? It is an unconventional approach to the problems of knowledge and knowing. It starts from the assumption that knowledge, no matter how it be defined, is in the heads of persons, and that the thinking subject has no alternative but to construct what he or she knows on the basis of his or her own experience. What we make of experience constitutes the only world we consciously live in. It can be sorted into many kinds, such as things, self, others, and so on. But all kinds of experience are essentially subjective, and though I may find reasons to believe that my experience may not be unlike yours, I have no way of knowing that it is the same. The experience and interpretation of language are no exception."

We will apply Glasersfeld's fundamental statement considering truth with respect to communication among individuals, however, in a mathematical sense. Before doing that some subjective impressions should motivate the work, as they involve the emotional perspective: Listening to music creates emotions. How come feelings and emotions when we watch musicians of different cultures, as shown in Figure 1?

H. Buchwald et al. (Eds.): S-BPM ONE, CCIS 85, pp. 24–33, 2010.

Fig. 1. Musicians with different cultural backgrounds

Fig. 2. The observer's filter when perceiving 'reality'

Each of them triggers emotions in our mind. But the emotion, a certain picture pro-voke, will be different for every single viewer, and this one is the only one the viewer can tell others. It represents the (individual) truth, and depends on the individual's so called "mental model". Now let us do an experiment: We are an external observer (with our individual mental models in the sense of experiences, knowledge, values etc.) to a group (organization) of individuals (see Figure 2). What do we recognize in the role of this external observer (could be as supplier or customer or colleague of another department)? We just want to see what we want to expect in sense of our truth. We do not expect how the group's truth is defined in the sense of the rules of their communication and processes. We expect that they fulfill our truth in terms of our expectations about their acting. But often these expectations are not met.

In order to explain these observations in a more concrete and practical way let us do an example based on the choreography of a ballet. The choreography of a ballet is defined by rules which have to be followed strictly:

In 1700 Raoul-Auger Feuillet defined five positions (rules). In this way he defined – as we say in our modern world – a method (see Figure 3 showing a notation of a Bourrée) how he could teach and publish these rules. But do we care about methods when we are watching a ballet? Of course not! We want to share the beauty of the dances in combination with the music, based on our personal experiences and expectations. Again, we define our own truth and hope the performance of the ballet will meet our expectations.

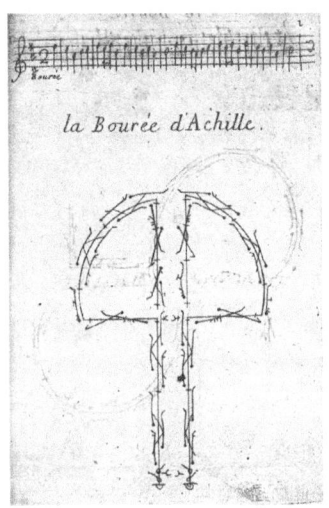

la Bourée d'Achille.

Fig. 3. Representing a Bourrée

Accordingly, we do not expect a static representation of a method as given in Figure 3, we rather expect a dynamic performance of actors implementing these methods, depending on the ideas of the choreographer, the interpretation of the music and of course, the personal behavior (see Figure 4). So the shared truth can only be a truth if the truth of all involved people is identical at a certain moment.

Similar examples could be found in literature, music and all kind of arts.

When generalizing these observations we can state: Everything that is perceived is individually interpretable. Hence, there are different, content-dependent 'realities' that result from individual interpretations. However, they are not mandatorily disjoint.

Fig. 4. Acting = Implementing a method

These realities are always real from the view of the respective individual interpreter. All of these realities are neither real nor wrong from the view of the external receiver of an interpretation. A view can only be exactly true if at a time t the individual interpretation is identical with the interpretation of the receiver of the interpretation.

Therefore, in a social context involving humans, a simple true/false dichotomy like in mathematics and informatics is too simple. What does this mean? Sending a verbal message to a human receiver means the message is always true from the sender's perspective. From the receiver's point of view it is neither true nor false. It would only be true for both if, at the same time t they share the same impression, experience, feelings and emotions. Then we think and feel in exactly the same way.

In business life we assume that we share the same impressions at the same time t, reducing communication and perception facts to a single dimension (money), in contradiction to the previous discussion. We deny that there are different expectations, different needs and values (see later, Figure 5). In general we deny that there are rapidly (i.e. every moment) changing truths', depending on the different actors involved in the shared process.

If we talk about global economic relations today, we primarily think of profit of global organizations acting effectively and efficiently. Such an understanding requires methods to observe and measure. They enable to speak a common language. We have learned in the course of the evolution that mathematics could aid the process of defining and implementing common languages. The reason is: Mathematics is objective by definition.

We simplify our world, since we assume: What is mathematically proved is objectively right (because the statements proven mathematically are always and everywhere valid). While applying these mathematical structures we rarely think about the approaches and the preconditions from the human observer's viewpoint.

An example: It is adequately known that the theorem of the Pythagoras is valid in a right-angled triangle namely:

$$a^2+b^2=c^2 \tag{1}$$

a,b describing the sides of the triangle and c its hypotenuse. This statement is always and everywhere valid as long as this is applied to a right-angled triangle. And the problem arises here:

If a triangle is defined without restrictions it is just a coincidence that a triangle happens to be a right-angled triangle. However, we often think we have per definition right-angled triangles when applying formula (1). In our economical life we (want to) reduce this truth to formulas with are likely simplified and do not represent a correct approach to the given situation.

Let us do another example based on the calculation of profitability: In economics the calculation of profitability is defined as

$$EW = \sum_{t=1}^{T} \frac{E_t}{(1-i)^t} \tag{2}$$

t being the index of period of profitability, EW the profitability, i an interest rate, and T the horizon of planning. As we mentioned above E_t is not a discrete number, it depends on so called "disturbing terms" based on behavior and communication of the interacting people like employees, suppliers and customers. These three categories have direct influence on the value of profitability. So we have to introduce the remainder term for series (2) in form of $\tilde{V}(t)$ which is defined as

$$\tilde{V}(t) := \begin{pmatrix} \tilde{v}_e(t) \\ \tilde{v}_s(t) \\ \tilde{v}_c(t) \end{pmatrix} \tag{3}$$

where $\tilde{v}_e(t)$ describes the behavior of the employees, $\tilde{v}_s(t)$ the behavior of the suppliers, and $\tilde{v}_c(t)$ the behavior of the customers in vector (3). But the elements of the vector itself are matrices defined as "behavior and communication matrices" below. Matrices in the sense that we can observe different behaviors (for example: behavior of payment of customers, absence caused by sickness of employees, behavior of payment of the suppliers, etc).

$$\tilde{v}_e(t) := \begin{pmatrix} \tilde{v}_{e_{11}}(t) & \cdot & \cdot & \cdot & \tilde{v}_{e_{1n}}(t) \\ & \cdot & & \cdot & \\ \cdot & & \cdot & & \cdot \\ & \cdot & & \cdot & \\ \tilde{v}_{e_{m1}}(t) & \cdot & \cdot & \cdot & \tilde{v}_{e_{mn}}(t) \end{pmatrix}, \tag{4}$$

$$\tilde{v}_s(t) := \begin{pmatrix} \tilde{v}_{s_{11}}(t) & \cdot & \cdot & \cdot & \tilde{v}_{s_{1n}}(t) \\ & \cdot & & & \cdot \\ & \cdot & & \cdot & \cdot \\ & \cdot & & & \cdot \\ & \cdot & & \cdot & \cdot \\ \tilde{v}_{s_{m1}}(t) & \cdot & \cdot & \cdot & \tilde{v}_{s_{mn}}(t) \end{pmatrix} \quad \text{and} \tag{5}$$

$$\tilde{v}_c(t) := \begin{pmatrix} \tilde{v}_{c_{11}}(t) & \cdot & \cdot & \cdot & \tilde{v}_{c_{1n}}(t) \\ & \cdot & & & \cdot \\ & \cdot & & \cdot & \cdot \\ & \cdot & & & \cdot \\ & \cdot & & \cdot & \cdot \\ \tilde{v}_{c_{m1}}(t) & \cdot & \cdot & \cdot & \tilde{v}_{c_{mn}}(t) \end{pmatrix}. \tag{6}$$

Based on this approach (2) has to be modified to

$$EW = \sum_{t=1}^{T} \frac{E_t + \tilde{V}(t)}{(1-i)^t} = \sum_{t=1}^{T} \frac{E_t}{(1-i)^t} + \sum_{t=1}^{T} \frac{\tilde{V}(t)}{(1-i)^t}. \tag{7}$$

About the remainder term

$$\sum_{t=1}^{T} \frac{\tilde{V}(t)}{(1-i)^t} \tag{8}$$

with the considerations (3) to (6) it is not possible to find any conclusion about its behavior because the functions in the matrices (4) to (6) are changing every time stamp t without any predictable or definable structure especially when we try to analyze the behavior with

$$\lim_{T \to \infty} \sum_{t=1}^{T} \frac{\tilde{V}(t)}{(1-i)^t}. \tag{9}$$

In our economical life we assume

$$\sum_{t=1}^{T} \frac{\tilde{V}(t)}{(1-i)^t} = 0 \tag{10}$$

and this situation is not only unsatisfying, but absolutely wrong. It can be true but it might not be true. So we just can conclude the truth of profitability a posteriori. A posteriori we know the empirical values of the "behavior and communication matrices" (4) to (6). There is no way to find a closed term of expression (10) because the time-depending complex structured function *can* change its behavior at every moment t, where the behavior of this function at the moment t can be completely different and independent from the behavior at the moment *t-1* or/and *t+1*.

Therefore we have to ask the question whether just the mathematical approach is the correct and complete one, even in regard to the dynamics we experience in our recent economical life. I clearly deny that. Maybe we can find another approach:

Einstein tells us: "The experience of an individual seems to be sorted in a series of experiences in which the single experiences in our retrospection are driven by the time of "earlier" and "later". So for the single individual there exists a so-called "Me-Time" or subjective time. This is essentially not measurable. Of course there is the possibility to assign number to the experiences so that the earlier experience gets a smaller number than the later one. But this is just arbitrarily. We can fix this assignment by definition of a watch where we can compare the flow of experiences with others." This means that at every moment t each individual has his/her own truth based on his/her experiences and expectations. With these considerations Einstein started his lessons about the theory of relatives. Thus, there is a possibility to use this famous theory to find valid mathematical models which take care in a correct sense describing the communication of individuals based on their single truths'.

Beside the mathematical considerations we have to take care that we can describe empirically the truths' in regard to our daily business life to reflect rapid changes and

Socio-Technical System

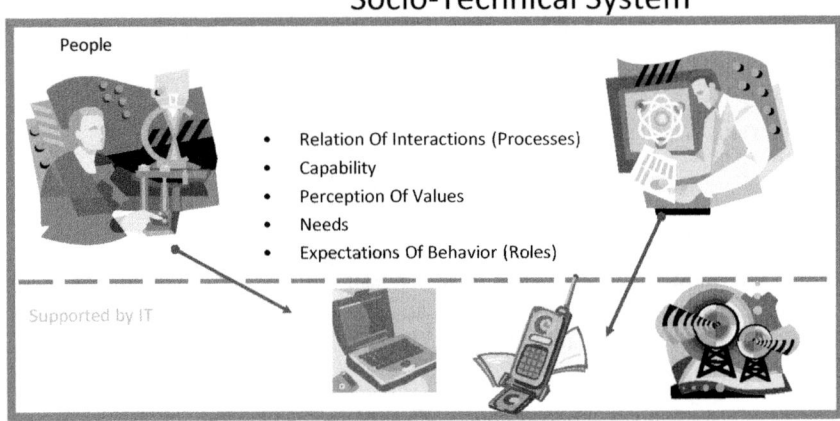

Fig. 5. Socio-technical system design

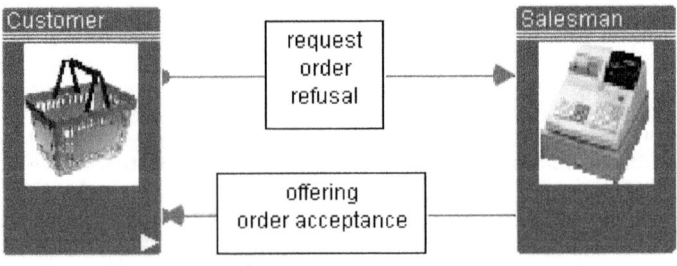

Fig. 6. Interacting subjects

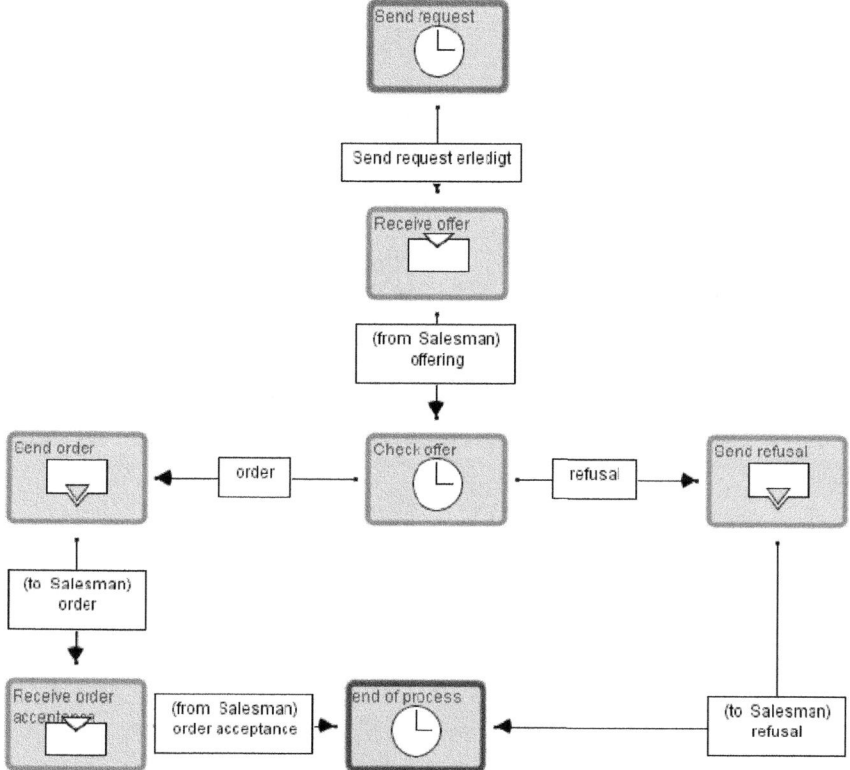

Fig. 7. Behavior specification of subject Customer

dynamical flows based on the individual expectations. In socio-technical system design the goal is to support these expectations with machines in form of IT systems. Behaviors of people have to change very quickly due to markets and customers. This means that the interaction between people in terms of communication has to change quickly and dynamically. Traditionally persons solve problems that could not described by mathematical formulas. So we have to find communication methods in addition which take care of all aspects of the socio-technical system where work processes occur (see Figure 5), taking into account

- relations of interactions,
- capabilities,
- perception of values,
- needs, and
- expectations.

Actually we need to care about direct interactions between people supported by information technology during execution time, rather than a posteriori.

Consequently, we have to focus on subjects who are handling the processes. We need to find a methodology to describe the processes in an intelligible way, in order to

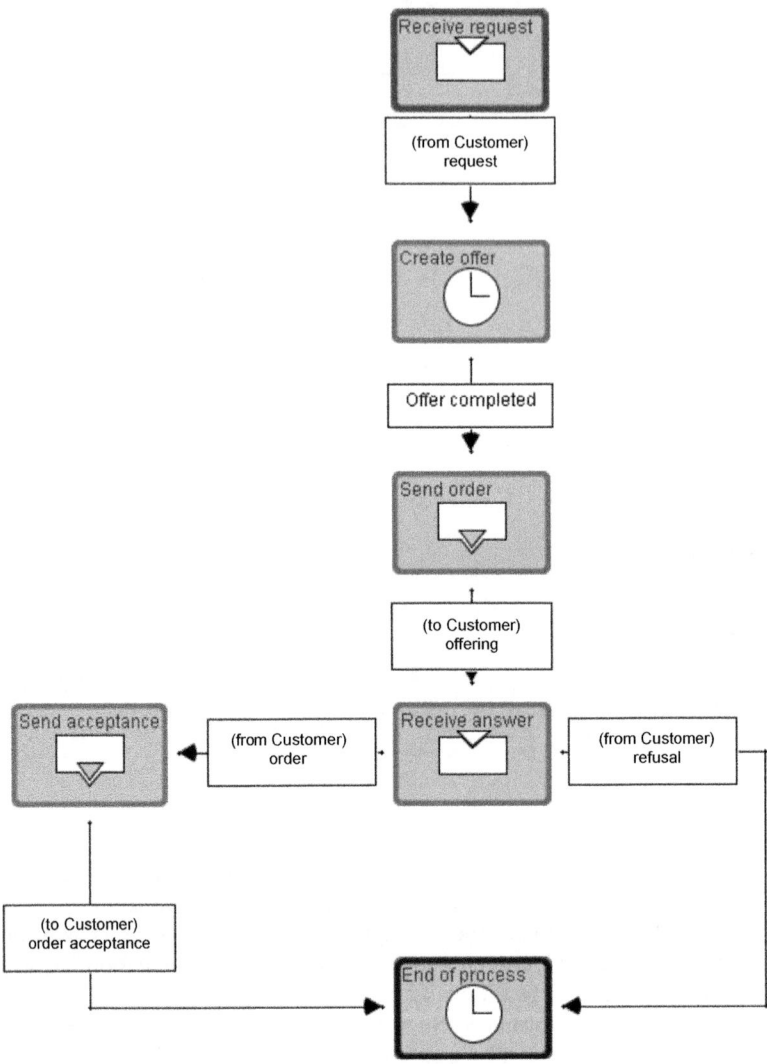

Fig. 8. Behavior specification of subject Salesman

empower the persons executing business processes. From the past we know the so called control flow-oriented methodology[1] which requires a lot of modeling experience to describe processes. In addition, these approaches have the deficiency of being not executable (cf. [6], [7]).

Fleischmann ([4], p.8ff.) has introduced the so-called subject-oriented description language. It allows describing the interchange of messages between subjects, and their

---

[1] And the object-oriented description language to describe IT-structures.

internal behavior. Figure 6 shows the interaction of 2 subjects and Figure 7 and 8 their internal behavior in terms of this diagrammatic language.

Complementary to the control–flow oriented approaches the subject-oriented method allows the immediate execution of the processes. In contrast to existing modeling approaches, the method is easy to understand, since it requires only 5 symbols to describe processes completely. The description of interactions between individuals is interpreted by IT systems in terms of "Do", "Send", or "Receive" a message that contains the necessary information (data) to be handled along the workflow. The subject – representing the daily work of each individual involved in a process – is in the focus of the process description. Therefore, the truth of each individual (in the sense of a single person or as a group of persons representing a common mental model) is mapped on subject representations that can be executed by computer systems.

The most important advantage of this method is that asynchronous processes can be synchronized, in the sense that individuals remain in control of acting at every moment t. Hence, we do not need to calculate the matrices (4) to (6) in a mathematical way. Actors find the solution in describing their processes and change them whenever they need, empowered with an immediate execution facility. In this way, in addition to mathematics we could find the missing link between individuals and computers representing individual truths.

References

1. Ballwieser, W.: Unternehmensbewertung – Prozess, Methoden und Probleme, vol. 2, Auflage, Stuttgart (2007)
2. Einstein, A. (1922): Grundzüge der Relativitätstheorie, vol. 7. Auflage, Heidelberg (2009)
3. Feuillet, R.-A. (1700): Chorégraphie, ou l'art de décrire la danse par caractères, figures et signes démonstratifs,
   http://de.wikipedia.org/wiki/Raoul-Auger_Feuillet (10.04.2009)
4. Fischer, H., Fleischmann, A., Obermeier, S.: Geschäftsprozesse realisieren – Ein praxisorientierter Leitfaden von der Strategie bis zur Implementierung, Wiesbaden (2006)
5. Hörmann, F., Haeseler, H.: Die Finanzkrise als Chance. Wien (2009)
6. Scheer, A.-W.: ARIS – Vom Geschäftsprozess zum Anwendungssystem, vol. 3. Auflage, Berlin (1998)
7. Schmelzer, H.J., Sesselmann, W.: Geschäftsprozessmanagement in der Praxis, vol. 4. Auflage, München (2004)
8. Von Glasersfeld, E.: Radical Constructivism – A Way of Knowing and Learning, London (1995)

Establishing an Informed S-BPM Community

Werner Schmidt[1] and Christian Stary[2]

[1] University of Applied Sciences Ingolstadt, Business Information Systems
Esplanade 10, 85049 Ingolstadt, Germany
Werner.Schmidt@haw-Ingolstadt.de
[2] Johannes Kepler University, Business Informatics,
Freistädterstraße 315, 4040 Linz, Austria
Christian.Stary@jku.at

Abstract. Modeling and executing business processes in a subject-oriented way can be considered as a paradigmatic shift in Business Process Modeling and Management. To be effective, traditional functional flow-oriented approaches have to be overlaid and superseded with patterns of actor interactions and the exchange of messages relevant for accomplishing tasks. However, subject-oriented Business Process Management (S-BPM) will only get substantial momentum once the mindset of organization designers and developers has started to change from thinking in terms of functions to thinking in terms of actor interaction. In this paper we propose an initiative designed to trigger that shift in a Community of Practice based on skillful education. In this way, revisiting traditional business process modeling and management can be coupled with experiencing alternatives effectively.

Keywords: Business Process Modeling, subject-orientation, Community of Practice, education, e-learning.

1 Introduction

'Insufficient sets of constructs, models, methods, and tools exist for accurately representing the business/technology environment. Highly abstract representations (e.g., analytical mathematical models) are criticized as having no relationship to real-world environments. On the other hand, many informal, descriptive IS [Information System] models lack an underlying theory base. The trade-offs between relevance and rigor are clearly problematic; finding representational techniques with an acceptable balance between the two is very difficult.' [8] Challenging this deficiency subject orientation is an emerging paradigm in Business Process Management (BPM). It provides balanced consideration of the actors in business processes (persons or systems are subjects), their actions (predicates), and their goals or the subject matter of their actions (objects). In addition, it allows straightforward implementation of specified processes. Organization developers, stakeholders, and managers can experience process specifications (cf. [5], [11]).

H. Buchwald et al. (Eds.): S-BPM ONE, CCIS 85, pp. 34–47, 2010.
© Springer-Verlag Berlin Heidelberg 2010

As Subject-oriented Business Process Management (S-BPM) is established as communication-based business development technique a community bringing together people interested in and working on S-BPM could serve developers and modelers and trigger S-BPM's diffusion in organizations, in particular, since other approaches to BPM are already established. The S-BPM community concept itself is based on Communities of Practices [13]. Communities of Practice (CoPs) can be established as informal social networks within or across institutional boundaries. Their members have similar interests, work on similar items or accomplish similar tasks, such as adopting a notation for subject-oriented business processes modeling.

The overall benefit of the S-BPM Community could be avoidance of misconceptions and diversification of developments, as it could be observed for UML developments (very late streamlining, lack of a meta-model reflecting its utilization etc.).

The S-BPM Community approach fits in and adds to an overall concept for establishing S-BPM as a new paradigm in BPM. The big picture consists of building blocks for technology (notation, architecture, reference implementation), methodology (patterns, process life cycle, maturity levels) and community (publication series, conference, community process) [2].

In the following we introduce the S-BPM community elements and their use along learning processes. In section 2.1 we describe the basic idea by giving an overview of elements and their context of use. Subsequently, the major elements are detailed. Section 2.2 is about the S-BPM WIKI. In section 2.3 we present the S-BPM.EDU environment as a shared learning space. In section 3 we provide a roadmap for implementation before we conclude reflecting the process in section 4.

2   The S-BPM Community – A Shared Learning Experience

2.1   Basic Idea

As shown in Figure 1 the central community element is the **S-BPM WIKI** which establishes a communication and interaction platform on the Internet for persons interested in S-BPM. Users of the wiki can be members of the Internet crowd. Their activities include generating, evaluating and improving all sorts of content around S-BPM. The wiki and its use in order to develop and explore content are discussed further in section 2.2.

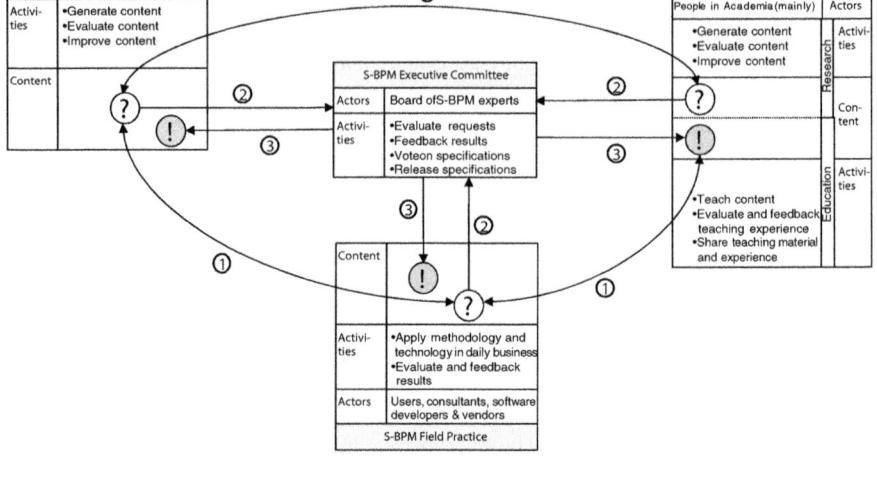

Fig. 1. S-BPM Community elements and process

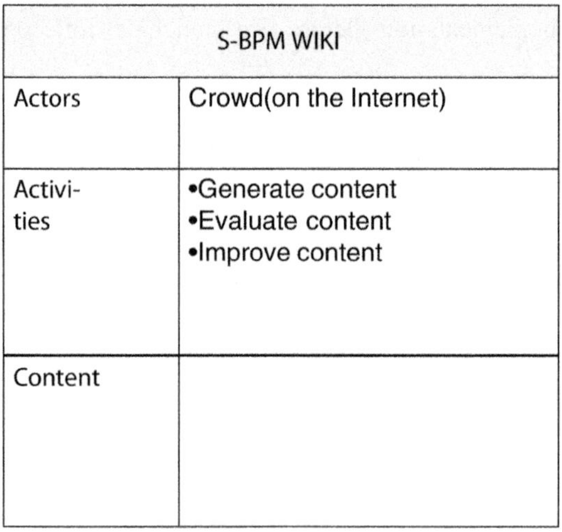

S-BPM WIKI	
Actors	Crowd(on the Internet)
Activi- ties	•Generate content •Evaluate content •Improve content
Content	

Fig. 1a. S-BPM Wiki

S-BPM Executive Committee	
Actors	Board of S-BPM experts
Activities	•Evaluate requests •Feedback results •Voteon specifications •Release specifications

Fig. 1b. S-BPM Executive Committee

S-BPM.EDU		
Actors		People in Academia(mainly)
Activities	Research	•Generate content •Evaluate content •Improve content
Content		
Activities	Education	•Teach content •Evaluate and feedback teaching experience •Share teaching material and experience

Fig. 1c. S-BPM.EDU

S-BPM Field Practice	
Content	
Activities	•Apply methodology and technology in daily business •Evaluate and feedback results
Actors	Users, consultants, software Developers & vendors

Fig. 1d. S-BPM Field Practice

S-BPM Field Practice allows business users and developers of the S-BPM methodology and technology to apply and evaluate ideas in a professional business setting, enabling immediate feedback and individual experiences (e.g. using the S-BPM WIKI).

The **S-BPM Executive Committee** could be an institution like the Java Community Process (JCP) Executive Committee, consisting of S-BPM experts from academia, business etc. The committee should define and control the community process leading from specification requests to specifications being published after having passed the voting (e.g., standards).

All elements are related by the **S-BPM Community Process** as indicated by the numbered arrows in the figure. In order to ensure this process' transparency and traceability, several questions have to be tackled, such as:

- When can a specification request be turned in to the Executive Committee?
  (Example: Publishing on the wiki and call for discussing it for a certain period of time)
- Who is eligible to turn in a request?
  (Example: A certain number of seriously interested community members)

In this way, the establishment of rules should become intelligible to interested persons and ensure long-term acceptance.

2.2 S-BPM WIKI

As a typical representative of social software wikis support human interactions in Communities of Practice. Providing functionality for easy posting, editing and commenting pieces of information as well as mechanisms for versioning and roll-back a wiki can serve as a powerful tool for managing knowledge about S-BPM.

Content can be posted and modified by accessing the wiki. Users can establish a topic they are interested in and actively develop it in cooperation with others. The so-called crowd generates content (user-generated content) and can also contribute by improving its quality (correctness, structure, level of covering and depth etc.). An interesting fact for exploring a new domain such as S-BPM is the incremental, evolutionary growth of wiki content. An author of a wiki entry can set a link to pages not yet existing. This mechanism allows identifying the need for further exploration and examination as well as complementing missing parts by other members of the community. For more details about wikis see [1] and [10].

In our context the S-BPM Wiki can be used to publish, publicly discuss and evolve issues considered to be relevant to S-BPM – from rough ideas to elaborated concepts. Contributors can be all sorts of members of the crowd, including researchers, practitioners, instructors, and students. Originally unstable, immature content can gain stability and maturity over time. If being considered to be sufficiently mature after several public revisions it can be turned in as a specification request to the S-BPM Executive Committee. Figure 2 shows a sample page of the S-BPM WIKI referring to basic modeling concepts.

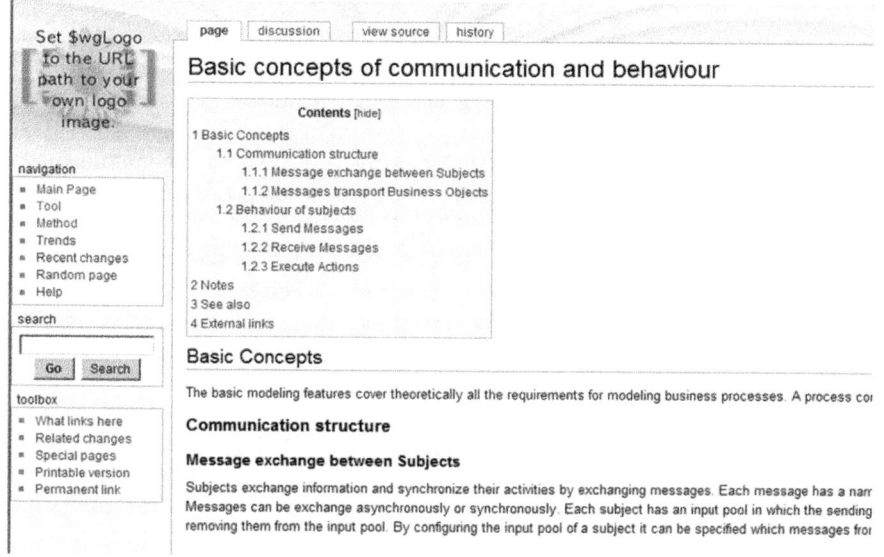

Fig. 2. Sample article on the S-BPM WIKI

2.3 S-BPM.EDU

2.3.1 Platform for Education and Research

The mission of S-BPM.EDU is to attract the academic community by building and providing an S-BPM platform supporting people in teaching, learning and doing research in the area of Subject-oriented BPM.

Figure 3 shows a selection of benefits and offerings for coaches and students as main target groups of the platform.

		
▪ Faculty	▪ Research environment	▪ Joint research projects ▪ Conferences ▪ Scholarships/Awards

▪ Students	▪ Teaching and learning environment	▪ Teaching and learning material
		  ▪ Curricula, case studies
		  ▪ Multimedia type learning units
		  ▪ Model repository
		  ▪ Demonstration and Training company
		▪ Variety of learning paths

Fig. 3. Major S-BPM.EDU target groups, benefits and possible offerings

As shown in the figure major benefit for faculty and students is having a sound teaching and learning environment. It could include teaching material like curricula proposals, case studies, model repositories, and a demonstration and training company. The environment should provide a variety of learning paths allowing learners flexible access to quality assured, stable S-BPM content, e.g., in form of multimedia type learning units. Instructors require authoring features to prepare and structure content in a didactically valid way (see section 2.3.2).

In the long run S-BPM.EDU could generate benefit for a target group not included in the figure. As it allows high quality education it helps the S-BPM eco system, composed of companies using S-BPM, consulting firms, software vendors and others (see S-BPM Field Practice), to identify and recruit well educated BPM experts in the future.

2.3.2 Didactic Design of the S-BPM.EDU Learning Space

2.3.2.1 Approach. For education purposes with respect to S-BPM we follow a bipartite approach. On the one hand individual learners and groups of learners are supported to build up S-BPM capabilities in a self-organized way. On the other hand, a curricular structure in terms of courses and modules allows learners to achieve various levels of competence. The virtual education environment needs to enable both, as S-BPM education follows the strict policy of a single point of reference for designing the structure for education and guiding the education process itself.

From the didactic perspective information sources have to be categorized in the collaborative learning spaces to support the active construction of knowledge. Active (re-)construction is seen as particularly beneficial for learners, since they can pursue their individual interests, while they are motivated to communicate their understanding to others. Such an approach reflects the situated and public nature of any construction activity [4]. The learning process can be guided accordingly by defining contracts as an active part of learning management [12]. Learner assignments for S-BPM education need to tackle both, the cognitive and the social aspects of coaching and learning in a mutually tuned way. However, their application for effective learning requires, besides the categorization of content elements and processing, dynamic link facilities establishing the communication context of when handling content [9].

We first introduce the community concept for learning, since it clarifies the social context of knowledge generation (section 2.3.2.2) before presenting learning features of the S-BPM.EDU environment (section 2.3.2.3) for situation-sensitive learning (section 2.3.2.4) and finally raising design issues of learning contracts (section 2.4).

2.3.2.2 Building Learning Communities. Learning communities can be regarded as a special representative of the Communities of Practice with characteristics already mentioned in section 1. The common interest of community members in the context of S-BPM.EDU is to learn about S-BPM, e.g. how to construct subject-oriented business processes. For this reason we talk of S-BPM.EDU Learning Communities of Practice (S-BPM.EDU LCoPs). According to the learning task and personal interest, membership of individuals in S-BPM.EDU LCoPs is fluent. Persons might switch between various S-BPM.EDU LCoPs. In this way, not only self-managed learning can be enabled, but also various settings and style of capacity building can be supported. Finally, as learners and coaches form peer groups for certain tasks, they might mutually benefit from different learning or problem-solving strategies.

S-BPM.EDU LCoP members might be regulated by their formal role, such as coach and learner, however, depending on the situative context, they might switch. Learners might guide others or elaborate a learning path they have explored for their coach. Coaches learn novel ways of exploring the content space and recognize information gaps that need to be filled for the next learning cycle.

In the mathetic S-BPM learning setting the coaches share the responsibility with the learner to accomplish commonly agreed learning tasks or goals. In this way, conform to traditional CoPs, members of the S-BPM.EDU LCoPs share perspectives, concerns or challenges for a certain period of time. Typical activities in S-BPM.EDU LCoPs are trainings for individuals or groups, conferences, thematic get-togethers, on-line news, and periodical exchange of experiences, techniques or ideas to solve problems or accomplish tasks.

Coaches and learners can benefit from S-BPM LCoPs in educational settings in several ways:

1. The reuse of knowledge is facilitated. Individuals might refer to relevant and shared documents within their peer group or with the coach.
2. Members mentor individuals and start discussions and discourse on S-BPM.
3. Members might enforce learning curves for newcomers, either through mentoring or referring to experts and contacts for particular themes. They might even facilitate the access to work practices and well-established methods, tools and procedures.
4. They serve as breeding force for novel ideas, techniques, and decision making.

However, coaches and learners need to structure learning arrangements and management. These should capture subject-specific elements that can easily become part of social interactions when learning or intervening in the learning process.

2.3.2.3   S-BPM.EDU Environment. The S-BPM.EDU environment is grounded on self-containing learning content tagged with didactic information [6]. It reflects the domain- and learning-relevant decomposition of information or material into so-called blocks. These blocks represent didactic information types, such as 'definition'. As they are also encoded into different media (text blocks, graphic elements, videos etc.) multiple (re)presentations of content may exist. Hyperlinks between blocks are common in e-learning to capture relationships between domain elements.

The S-BPM.EDU environment allows different levels of detail (LOD) for each block, e.g., providing a slide for a definition on the top level (LOD 1) based on the full text of the definition on LOD 2 (representing a textbook). Annotations (see also marked feature bar in Figure 4) constitute individual views on content items by commenting, linking, or highlighting item block elements, or enriching content blocks [7]. One of these annotations could be links to communication entries of the S-BPM.EDU environment communication components (chat, forum, infoboard etc.). In this way communication elements are directly linked to content blocks and *vice versa* (middle bottom to right in the Figure 4). Communication needs to be established among peers for learning, as well as between learners and coaches. The latter, in the role of quality managers, are responsible for improving content and settings based on learner input and feedback.

The content is arranged according to the aforementioned information types (see bottom of hierarchy in the figure). Currently, 15 generic types of this sort are available as part of an XML scheme. It comprises definition, motivation, background, directive, example, self test, and other didactically relevant content structures. Some domain-specific block types, such as proof for math content, have been added to support domain-specific applications. Each block type can be visualized in S-BPM.EDU through a certain layout, e.g., colored background.

2.3.2.4   Situation-Sensitive Learning. Block types allow learners to scan the entire learning content for specific categories of information using a filter function. The workspace then shows only selected block types. In this way learners might follow their situation, interests and habits, such as starting to learn with studying background information.

Another typical example is the preparation for formal exams. Students might select examples or definitions when starting to learn.

The bar on top of the content area in Figure 4 allows for annotating the information of a learning unit according to individual needs. The users might switch between various levels of detail, using the LOD function besides the filter in the function bar. With respect to content manipulation, in addition to marking content elements users might link blocks to internal or Internet-based sources of information, as well as to entries of a discussion forum or other S-BPM.EDU communication tools (chat, blog, a.t.l.). All annotation activities are stored in user-specific views that might be shared and cascaded, using the view functionality located on the utmost left side in the function bar.

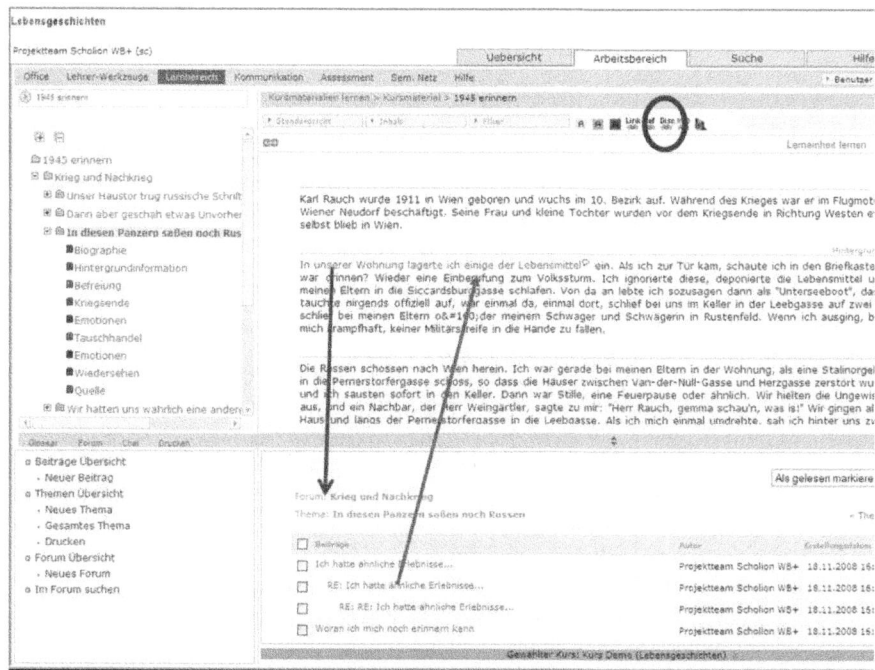

Fig. 4. Content elements can be directly linked to communication entries (and vice versa)

Figure 4 shows the direct link between content elements and communication entries of a forum. In the content area this link is indicated with a speech bubble, in order to distinguish it from traditional links to other content elements. Users set those links, in order to keep the history of conversation with their partners. Being part of asynchronous communication (here using a forum) users might interact whenever convenient for them. Due to the integrated handling of content and communication in S-BPM.EDU, content individualization as well as communication (entries eventually linked directly to content items) becomes traceable without losing the original context.

2.3.2.5 Guiding Individual Learning Processes. Following the tradition of self-managed learning, mathetic components need to be provided to guide learners through exploration of content and problem solving procedures. We use Intelligibility Catchers (ICs) [12] as they are e-learning assignments designed to facilitate individual capacity building and to establish sustainable learning communities. The first objective is achieved by providing content-related orientation and work tasks. The latter is achieved by sharing views on content elements and (intermediate) results in a discursive way. ICs embed the e-learning features in learning activities. Table 1 exemplifies the promotion of theoretical concepts to help understand subject-oriented modeling and specification.

In the *orientation* section the stage of capacity building the IC should be used for is addressed, and what learners can expect when accomplishing the IC tasks. The *objectives* set the scope in terms of the topics that are addressed and the understanding that

Table 1. A sample S-BPM.EDU Intelligibility Catcher

1 – Preface / Orientation	Modeling is a core activity in organization design, system and software development. So far the demand for modeling has been motivated. Now a specific modeling technique is introduced. The assignment helps exploring S-BPM, a modeling language for business processes.
2 – Objectives	Understand S-BPM from a modeling perspective, including the S-BPM rationale.
3 – Tasks	- Capture the theoretical background of S-BPM - Apply S-BPM for distributed work tasks - Discuss your results with peers in the respective discussion forum
3.a Documented Work	• *Filter* content for 'background information' • Develop *view*[1] ,Theory.x' for each theoretical concept found in the prepared content • *Search* for additional information or original web source for each theoretical concept of S-BPM – once you have found a reliable source of information, set a corresponding link in the view. • Search for typical instances of each concept in terms of S-BPM diagrams. Each time you have found one, *annotate* each diagram identified in the content area with an example of your choice. • Make views *public* to other peers and the coach • Describe your results in dedicated linked entries of the *discussion* forum • Compare and reflect results in topic-specific *chats*
3.b Intellectual Challenge	• (Re-)Construction of material • Develop individual position
4 – Conferences	Continuous feedback by peers and coaches

---

[1] A view in the S-BPM.EDU learning environment is a virtual overhead slide put on top of content items containing all annotations. It can be made public to share annotations with other users, or remain private for further processing.

Table 1. (*continued*)

5 – References	http://www.jcom1.com
6 – Bulletins	Infoboard@S-BPM.EDU
7 – Departmental Cuts	This assignment should take you no longer than 20 hours.

should result from exploring and processing topic information. It reflects the mathetic value of that learning unit. The *task* section comprises a documented and an intellectual work part. It encourages active information search and processing, communication, and personality development. The *conference* section sets the rules for the community of practice. The *reference* section provides links to material that helps to accomplish the tasks. The *bulletins* can be dynamically created and are available in the infoboard. Finally, the *departmental cuts* reveal the estimated individual effort to meet the objectives.

The structure combines organizational with subject-specific information arranged from a mathetic perspective. For instance, the orientation section in the beginning informs coaches and learners when to use this IC addresses competencies, the content involved, and the rationale for exploring content and co-constructing mental representations. Initially, the learners are encouraged to identify those blocks of the learning unit where concepts are already documented, i.e. part of the prepared content. Then they are asked to complement particular content items, namely S-BPM diagrams, having no information about their origin so far. After practical modeling, all results should be shared with peers, enabled by dedicated views and focused, since these results represent content-related discussion items. All results are validated by the coach through feedback, in order to ensure correct learner representations.

Reference points for exploration and communication are the content types previously addressed. The block types can be arranged according to access or learning patterns, such as a motivation serving as entry point to an explanation, followed by a definition or an example, learning style and situational preference [3]. ICs enable explicit learner control, as they reveal the variety of paths to be followed for knowledge generation.

Coaches can utilize this kind of representation in a variety of ways. On the one hand, they might use such a scheme to reflect their own style of preparing material and their ways of intervening along knowledge generation processes. On the other hand, they might use it as a tool to compare existing approaches to domain-relevant materials, such as existing textbooks, hypermedia or other study material.

Finally, the contract specification benefits from such a structured input. Given didactic guidelines, coaches can directly apply them when preparing or arranging learning material. Block types, ontologies and learning paths are design elements that allow coaches the explicit implementation of guidance according to their understanding and degree of intervention. They can not only use explicit didactic information for preparing material, but also when suggesting alternative learning paths for learners needing exploration support in self-organized learning settings.

3   Roadmap of Developments

So far, the stage has been set providing all ingredients to build a community for educating organization designers and developers in a skilful way. The next steps focus on adapting existing, if not generating S-BPM material for education, recruiting community members, and starting a Community of Practice. In parallel, continuous evaluation and feedback checks have to be established, in order to tune modeler needs and product developments.

- *Material development*: BPM is subject to academic teaching and occupational training. Existing material needs to be reviewed with respect to background and conceptual information that is required to convey the different capabilities of approaches.
- *Creating awareness*: In parallel to the development of educational material existing communities in research and practice need to be approached to share the reflection of existing BPM techniques. Typical offerings in this respect comprise meta-modeling trainings, as meta models do not only show the structure of BPM notations but also their handling and context of use.
- *Starting a Community of Practice*: This initialization allows business users and developers of the S-BPM methodology and technology to apply and evaluate educational ideas in business settings and to gather educational requirements. Role assignments, structure elaboration, and netiquette building are the most important tasks to start a constructive interaction process.
- *Establishing continuous feedback*: As the envisioned approach does not only involve academic but also industrial educators, mutual feedback loops need to be established. They allow practitioners to work with the material and learning technology and share their experiences directly with developers.
- *Monitoring and guiding the process*: The S-BPM Executive Committee needs to establish a sustainable organizational and economical infrastructure for the CoP. It should consist of S-BPM experts from academia, business, standardization bodies, and organizational and tool development communities.

4   Conclusions

Recognizing the proliferation of subject-orientation as a paradigmatic shift in Business Process Modeling and Management requires tackling the mindset of organization designers and developers. They do not only have to rethink existing approaches to acquiring, representing and executing business processes in terms of functional workflows, but rather have to understand business operations as patterns of actor interactions and the exchange of messages relevant for accomplishing tasks.

In this paper we have proposed an initiative designed to trigger that shift. We have suggested community building based on skillful education. Central to the community is the **S-BPM WIKI**, as it allows establishing a communication and interaction platform on the Internet for persons interested in S-BPM. Their activities include generating, evaluating and improving all sorts of content around S-BPM. **S-BPM.EDU** serves as an environment for education and research with respect to S-BPM. It should support researchers developing knowledge in the area, and provide means for coaches to build up S-BPM capacity by making use of didactic S-BPM entities. The **S-BPM Field Practice** allows business users and developers of the S-BPM methodology and technology to apply and evaluate ideas in a professional business setting, enabling immediate feedback and individual experiences (which should be available through the S-BPM WIKI).

The **S-BPM Executive Committee** comprises S-BPM experts from academia, business and affine S-BPM communities. Its task is to define and control the community process leading from specification requests to specifications being published after having passed the voting. The resulting **S-BPM Community Process** needs to be transparent and traceable. In this way, revisiting of traditional business-process modeling and management can be coupled with experiencing alternatives effectively.

References

1. Back, A., Gronau, N., Tochtermann, K.: Web 2.0 in der Unternehmenspraxis, Oldenbourg, München (2008)
2. Buchwald, H.: Potential Building Blocks of S-BPM. In: Buchwald, H., et al. (eds.) S-BPM ONE. CCIS, vol. 85, pp. 123–135. Springer, Heidelberg (2010)
3. Euler, D.: Forschendes Lernen. Universität und Persönlichkeitsentwicklung, Campus, Franfurt/Main (2005)
4. Farmer, R.A., Hughes, B.: A Situated Learning Perspective on Learning Object Design. In: Proc. ICALT 2005. IEEE, Los Alamitos (2005)
5. Fleischmann, A.: What is S-BPM? In: Buchwald, H., et al. (eds.) S-BPM ONE. CCIS, vol. 85, pp. 85–107. Springer, Heidelberg (2010)
6. Friesen, N.: Three Objections to Learning Objects and E-Learning Standards. In: McGreal, R. (ed.) Online Education Using Learning Objects, pp. 59–70. Routledge, London (2004)
7. Fürlinger, S., Auinger, A., Stary, C.: Interactive Annotations in Web-based Learning Environments. In: Proc. ICALT 2004, pp. 360–364. IEEE, Los Alamitos (2004)
8. Hevner, A.R., March, S.T., Park, J., Ram, S.: Design Science in Information Systems Research. MIS Quarterly 28(1), 75–105 (2004)
9. Kienle, A., Herrmann, T.: Integration von Kommunikation und Kooperation an Hand von Lernmaterial – ein Leitbild für die Funktionalität kollaborativer Lernumgebungen. In: Proc. Mensch & Computer 2002, GI & ACM German Chapter, Teubner, Stuttgart, pp. S45–S54 (2002)
10. Koch, M., Richter, A.: Enterprise 2.0, Oldenbourg, München (2007)
11. Schmidt, W., Fleischmann, A., Gilbert, O.: Subjektorientiertes Geschäftsprozessmanagement, HMD – Praxis der Wirtschaftsinformatik, vol. 266, pp. 52–62 (2009)
12. Stary, C.: Intelligibility Catchers for Self-Managed Knowledge Transfer. In: Proc. ICALT 2007. IEEE, Los Alamitos (2007)
13. Wenger, E., McDermott, R., Snyder, W.: Cultivating Communities of Practice – a Guide to Managing Knowledge. Harvard Business School Press, Boston (2002)

Business Process Management – S-BPM a New Paradigm for Competitive Advantage?

Robert Singer and Erwin Zinser

FH JOANNEUM – University of Applied Sciences, Austria
{robert.singer, erwin.zinser}@fh-joanneum.at

Abstract. This article summarizes our motivation to deal with business process management in several dimensions. Firstly, a critical survey on the current status of business process management in industry is given. Based on well known facts from the area of business administration and strategy, we show why business process management as a whole needs to renew its paradigm, especially in the context of IT support. We further demonstrate that one emerging methodology, the notion of subject oriented business process modeling, is one of the promising candidates to change the way of working and to unfold the full potential of business process thinking in organizations. Secondly, we provide some insights into current educational and research agenda at our bachelor and master program Information Management, respectively, considering the subject oriented business process modeling approach as a valuable alternative to heretofore established procedure models. In conclusion, a brief outlook on future actions concerning S-BPM in research and education at our department is given. Business Process Managgement

1 Why Do We Need Business Process Management?

1.1 Motivation for Business Process Management

Inspired by the striking phrase "Does IT matter?" by Nicholas G. Carr [1], we would like to start our discussion with the question "Do we need business process management (BPM)?". We have the intention to start a critical review and discussion about the concept and the use of business processes in the context of profit and non-profit organizations.

Working together with about a hundred of companies over many years, sized from SME up to globally acting industry leaders, we are able to say, that various companies have programs for business process management in place, but it seems many of them often do not even know why or in best case they do not unfold the full potential of business process thinking as documented in literature (see for example [2] and [3]).

Many of these companies also have certificates which prove the compliance with industry standards (e.g. ISO 900x series) or they were initiated to implement complex (IT-)systems (e.g. Enterprise Resource Planning software such as SAP or Microsoft Dynamics). Mostly the compliance is achieved by establishing well documented business processes, which are audited by an independent

H. Buchwald et al. (Eds.): S-BPM ONE, CCIS 85, pp. 48–70, 2010.

trusted organizations. After that, the whole business process initiative goes into sleep mode up to the time of re-certification. Summing up, such business process management (BPM) arrangements are not really relevant for actual business needs.

This can also be emphasized by the fact, that many (maybe even most) of such initiatives were started by middle management because of some operational needs - often the strategic aspects (but to reach a certificate) and responsibilities are entirely missing. These are observations missing a scientific foundation based on reputable empirical studies, but let us use them as the starting point for a discussion about the original intentions of business process management, recent developments in the field, and a comparison of actual and future BPM needs of companies.

Although there have been earlier attempts to implement process thinking [4], a good starting point can be found in the prevalent book "Reengineering the Corporation" by Hammer & Champy [2]. The questions is: what happened with the ideas and the essence of BPM presented by Hammer & Champy?

Let us begin the discussion with a simple, clear, and straightforward definition of (business) processes as given by [3]: "a process is the way in which the abstract goal of putting customers first gets turned into its practical consequences. Without process, companies decay into a spiral of chaos and internal conflict". Therefore we should not forget, that customers are not at all interested in the activities towards which companies devote most of their managerial energies. Customers care about one thing only: results.

The problem is, that work creating results for customers often is still broken into pieces and scattered across numerous departments and units. People and departments focus on each of the steps that lead to creating results for customers, but no one focuses on all steps together as a unit - in best case this leads to sub-optimization, in worst case it results in customer dissatisfaction. The question is: why do we still work according to the paradigm of Taylor [6] and Smith [7]? In our opinion the apparent advantage of a process organization is widely accepted, the problem is based on the realization and this again has to do with the representation and execution of processes within information technology platforms and systems. The strong link of information technology and processes was worked out by Davenport in his well known Book "Process Innovation" [8].

As stated by Carr [1], there is no existing proof, that investment in IT leads to better results at the bottom line, but investment in excellent business processes does. As stated by Smith & Howard [5] it is the task of IT to support business processes, or even excellent business processes today are seldom possible without innovative support from IT.

Process is a word which is widely but often incorrectly used in the business world. Put most simply, processes are what create the results that a company delivers to its customers. Process is a technical term with a precise definition[1] [3]: 'an organized group of related activities that together create a result of value

---

[1] A very similar definition from operations management literature would be: 'A business process is a network of connected activities and buffers with well defined boundaries and precedence relationships, which utilize resources to transform inputs into outputs with the purpose of satisfying customer requirements'.

to customers'. While each manager makes sure that his or her department excels at its narrow task, no one ensures the excellence of the whole operation; nor does anyone view fulfillment as a whole through the prism of process.

Without precise process designs and common integrating goals, employees have little chance of consistently operating in ways that customers find convenient. They will even have less chance of successfully performing and coordinating the broader range of activities needed to deliver higher levels of value-added. As work gets more demanding and more complex, process becomes absolutely essential.

The above definition of the term process focuses on two words - *organized* and *together*. Being organized means having concrete, specific designs for processes so that their performance isn't determined by improvisation or luck. This is a core element of process management and contains the application of the Deming Circle (plan-do-check-act). Being together means creating an environment in which all process workers are aligned around common goals and see themselves as collaborators rather than antagonists.

Another aspect of business processes are their relation with the company strategy. To make it short and precise: processes are (only) means to bring the strategy to life and therefore to realize competitive advantage[2]. We are sure, that most readers will agree up to this point to our discussion.

1.2 Business Processes and Information Technology

Why is it so difficult to bring business processes up to their full potential? First, we have to realize that there are other viewpoints, as illuminated by Weske [10]. He remarks that business process management has received considerable attention by both business administration and computer science community. In other words, we are in the middle of the so called business-IT-gap. In general, many scientists are aware of the problem, but we do not think that a (final) solution is ready for applications yet.

We can find a myriad of books about how to conduct business process management, how to design (often only how to draw) business processes and so on. A search on amazon.com with the phrase 'business process management' gives us a list with much more than 20.000 books. We think, the main concept of business processes can be easily explained and the books about 'why companies should be process centric' should have already been written. The remaining problem can be expressed by having a look at the process life cycle as explained for example in [10] or [16]. Processes are not only analyzed, and designed, but they are also enacted. Enactment nowadays is mostly done with the help of IT support. In order to unveil the full potential of business processes we have to manage the whole business process life cycle where business and IT agenda have to be considered likewise.

---

[2] Here we use the term 'competitive advantage' intentionally to emphasize the dependency of business processes on the value chain as discussed by Porter [9].

We cannot discuss business processes without talking about IT-systems. Nowadays process support is more or less one of the core purposes of modern IT-architectures. Of course we have a broad spectrum of IT support ranging from ad-hoc human interaction workflows up to fully automated production workflows. As explained before, the main purpose of business processes is to realize a value chain and to overcome the paradigms of Taylor and Smith, which do not fit very well with our customer oriented and competitive environment.

What are the research efforts to overcome these obstacles up to now? If we have a look at past and recent scientific literature about processes, we find a lot of discussions for example about workflow patterns[3] [12], petri-nets [13], workflow-nets [14], as well as research on [11] process description notations (BPMN[4]) and process execution languages (WS-BPEL[5]).

Of course, the findings and analysis related to the so called workflow patterns are possibly interesting and important from an academic point of view and can very well be used as selling arguments for software vendors (they can argue who is supporting which pattern, or not), but do they really help companies to achieve competitive advantage implementing agile and adaptive business processes?

Another more or less technical theme is the "war" of process description standards. For the moment we take it for granted that BPMN is the first choice to "document" business processes. We do not believe, that a process documented with BPMN can be designed or even understood by people who are working in a process. Almost every book on business process management mentions exactly that as good practice. As a result a company (again) needs external or internal consultants who analyze, design and document the areas (the processes) which normally are called important assets of a company. Nobody should wonder, if those seemingly perfectly drawn processes are not adapted towards actual needs, or even not used in daily business. Hence, doing this has no economic value for companies - but raises costs only.

The underlying problem of bringing processes towards execution in an IT environment is even more complex and we follow [10] to explain it shortly here (see Fig. 1). Models are expressed in metamodels that are associated with notations, typically of graphical nature. For instance, the Petri net metamodel defines Petri nets (places and transitions) as a directed bipartite graph. The traditional Petri net notation associates graphical symbols with meta-model elements. It is important to distinguish between the concepts of a modeling approach, and the graphical notation used to represent these concepts. The complete set of concepts and associations between concepts is called meta-model. A notation associated with a metamodel allows expressing the concepts of that particular metamodel. All generally used graphical process notations follow this general

---

[3] A comprehensive bibliography can be found on http://www.workflowpatterns.com
[4] Business Process Modeling Notation, standard can be downloaded under http://www.omg.org
[5] Business Process Execution Language, standard can be downloaded under http://www.oasis-open.org

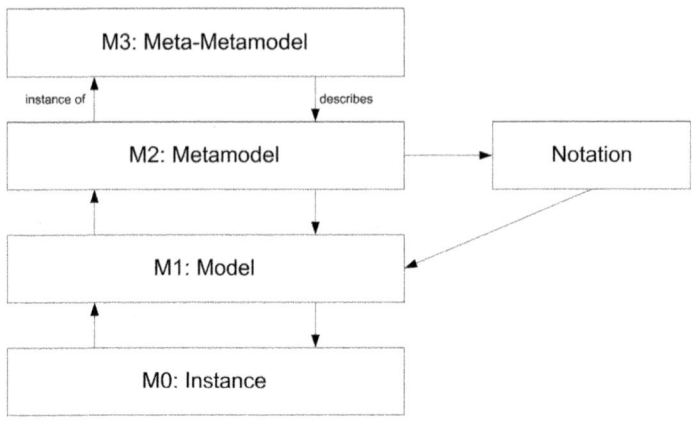

Fig. 1. Levels of Abstraction (Meta Object Facility, MOF)[15], adapted from [10]

concept. That means we use a certain meta-model (UML, eEPC, BPMN (2.0), Petri-net, ...) and the associated notation to document processes.

There are some pitfalls and drawbacks using this concept. Designing business processes is at first a top-down process - we have to implement the goals defined in the company strategy; contrary, if we analyze processes we start to work bottom-up first and afterwards we normally redesign or adapt the newly documented processes according to some predefined targets. Nevertheless we are always confronted with the question of how detailed the process should be and in the first stage we normally work out process descriptions in natural language. This text, has to be translated first into a meta-model of choice and afterwards with the help of the associated notation into a graphical representation of the process. These transformations[6] are time consuming and therefore costly and we often need - as mentioned before - "meta-model and notation experts" for this translation; obviously we would need the same experts to translate and interpret the documented processes back for the employees to work according to them. Additionally we normally lose information during this transformation process - as we make the "real processes" to fit the possibilities of the graphical notation. At the end we are confronted with the situation, that the owner of a business processes has lost the control of his or her process towards some technocrats, e.g. IT-staff, Quality Manager, etc.

---

[6] Additionally we are confronted with the so called "functional decomposition", i.e. the break down of detailed process descriptions from strategic value chains down into instantiated process pieces. At each level we are confronted with the question to define a certain level of detail.

1.3   New Insights and Mobile Processes

After reviewing the overall situation, we want to have a look towards innovative and promising ways to overcome the mentioned drawbacks of actual business process systems – and we think such concepts actually exist.

Besides practical aspects of real world applications of business processes and their representation in IT-Systems, we are confronted with an interesting academic discussion. This discussion obviously is initiated by the facts already stated in this paper, but – again – mostly has a strong technical and academic background. But we think, that at the end there can be innovation over the whole business process management life cycle. The terms discussed are lambda-[7] and π-calculus.

First, Smith & Fingar have discussed the implications of π-calculus based on economic aspects in [17][18][19], but they included technical aspects too (they particularly argued about BPML[8] as technical implementation of π-calculus for process definition and execution). A thorough commented literature analysis about π-calculus can be found in [21] and as an introduction in the seminal book of Milner [22].

Smith & Fingar explain the principles of π-calculus (which is the mobility of processes) with the help of the well known concept of email [19]: Consider electronic mail as a process. We can send an email to another person, this one, for example, forwards the email to a third party and this one is then able to communicate and collaborate with me as initiator of the email. How does this happen? By receiving email, or more specifically by receiving an email address, directly or indirectly, interchange the capability to communicate with others linked to that email address. This is what makes email work. We give a name, in the form of an email address, to others, and this gives them the ability to communicate with yet other participants in the thread of the conversation - continuously extending the conversation over time, involving new participants that contribute value to the process. Through this simple model a dynamic way of conversation becomes possible – a new business process.

Now, what's really interesting is that even this simple process cannot be easily modeled, and then executed, using classical workflow engines. The reason is that email processes exhibit so-called mobile behavior. Mobility is possibly a property of most, perhaps all, processes - a phenomenon recognized by Robin Milner [23]. This had been an active area of research prior to the identification of the π-calculus, both by Milner in respect of his earlier CCS (Calculus for Communicating Systems) and the work of others, including Anthony Hoare's process algebra for CSP (Communicating Sequential Processes) [24].

By adopting this approach to process representation, arbitrary distinctions between what is considered communication, and what is considered computation,

---

[7] In mathematical logic and computer science, lambda calculus, is a formal system for function definition, function application and recursion.

[8] Business Process Modeling Language; 2005 responsibility transferred to the OMG (`http://www.omg.org`) and included in the OMG-MDA-Modeling-Stack[20].

begin to dissolve. It's necessary to realize that in the world of π-calculus, all participants in a process are themselves processes.

In the domain of π-calculus, every process participant is given a unique name, and that name is a central notion of π-calculus – the connections between named participants represent the dynamic capabilities and behavior of any given process, at any point in time. π-calculus is an algebra in which names represent channels that can act both as transmission medium and as transmitted data. This communication is done on complementary (input and output) channels. The contents of messages are also channels. As a result of such a communication event, the recipient process may now use the received channel for further communication, as in the email example. This feature, the mobility in the system, allows the network "wiring" to change with interaction between the participants. Milner has shown that, mathematically, all that we previously understood as computation, and all that we previously understood as communication, can be modeled and understood as the same thing - processes [19].

The question now is, do we need mobility, or can we even draw further benefits out of this new way of thinking? And without any doubt, the π-calculus can be used to invent IT-systems which overcome all (or some) of the drawbacks emerged during last two decades. Actual research confirms the need of a new theoretical fundament of business process representation and execution and that π-calculus is a promising foundation for this [25]. As research, e.g. [26], in the domain of computer science is important, we should not forget to find economically useful applications to get the promised competitive advantage from business processes.

As we are in a scientific discussion we also should be aware, that there are mature arguments, that π-calculus is not really needed and actual theoretical concepts are sufficient to support the whole business process life cycle, which is for example pointed out by van der Aalst in his answer [27] to Smith & Fingar [19]:

> ..., I would like to emphasize that Pi calculus, as Robin Milner and others have developed it, is a solid foundation for modeling and analyzing processes. However, I'm not convinced that the features present in Pi calculus are vital for Business Process Management (BPM) systems. The only thing which distinguishes Pi calculus from classical process algebras like CCS, CSP, and ACP is the notion of mobility. For some application areas, this feature is very useful. However, for BPM solutions it seems less relevant; anything that can be expressed in terms of Pi calculus can also be expressed in other process algebras (extended with some notion of data) and other process models like, for example, high-level Petri nets. In any case, there is no clear evidence that Pi calculus supports patterns in a better way than more traditional languages like Petri nets.

1.4   Subject-Oriented Business Process Management (S-BPM)

The weak point of all the new paradigms is, that we need a proof of concept. Especially a proof of the added economical benefit in case of application and implementation. For that reason, we did some preliminary studies with a new commercially available software tool – the jCOM1 BPM Suite[9]. Some of these research projects are discussed in Part II of this paper.

This tool implements the so called subject-oriented approach towards business process management. The subject-oriented methodology [16] is based upon a process algebra for the modeling of parallel processes with subjects, elementary actions, and communication relationships as they were originally introduced in the 1980s through the informatics theories of Milner and Hoare. The theoretical concepts of discussed implementation (jCOM1) are laid down in [28]. A process designed in the domain of S-BPM consists of two layers: a communication layer, which contains the messages between subjects, and a behavior layer, which contains the internal behavior of the subjects.

The S-BPM approach is a promising step forward in BPM mainly because of the following facts:

- an integrated message orientation (this e.g. fulfills one of the requirements for new BPM as discussed by [25] 'A formalism representing the first shift of BPM should be based on messages, or events, rather than states')
- a behavior oriented modeling approach (this is also discussed in [26] in the case of π-calculus)
- a puristic set of graphical symbols – four symbols are sufficient to model all possible business processes; most other approaches (notations) need a lot more symbols[10]
- processes can be modeled with natural language in mind – that is the reason of the name of this approach: the acting parties in a process are called subjects, subjects can do something (predicate), and they can work on or exchange (business) objects, i.e. subject - predicate - object (a full sentence)
- process models are inherently strictly formal defined and therefore can be executed without any human intervention resp. adaption, that means no explicit (and additional) transformation is needed.

S-BPM and the reviewed implementation is a try to bridge the business-IT-gap. As mentioned before we think, that the possibility to think rather in natural language than in a meta-model and graphical notations should improve the design phase. Of course empirical studies are needed to prove the benefit of the approach. Furthermore – as discussed in the following chapters – technical questions are remaining too. As currently only one commercial realization is available, no comparisons between different tools (and their implementation philosophies) are possible.

Nevertheless S-BPM has considerable benefits from a didactically point of view to bring business process management closer to the people involved, especially through the possibility of immediate execution (testing) of the designed

---

[9] jCOM1 AG, Lilienthalstraße 17, 85296 Rohrbach, Germany; http://www.jcom1.com
[10] e.g. BPMN 1.1 consists of 52 distinct graphical elements: 41 flow objects, 6 connecting objects, 2 grouping objects, and 3 artifacts.

processes by the involved people[11]. In addition design-cycles can be drastically shortened which gives a clear cost benefit [29].

2   S-BPM in Education and Research

In order to introduce S-BPM into our degree program Information Management, we have been taken two approaches: Firstly, students should be stimulated to study and critically evaluate the subject-oriented design paradigm for business processes in comparison to heretofore state-of-the-art procedures, e.g. BPMN swim lanes or ARIS EPCs. They should get familiar with more intuitive and human-like thinking as well as procedures resembling "the way of working" with a quite clear objective of learning to design and execute business processes in an understandable way. Secondly, we were eager to investigate S-BPM from the scientific point of view with strong emphasis on collaboration (e.g., unified communications) and human-interaction activity, communication and collaboration patterns.

The following sections deal with representative topics from our research group as successful examples for the adoption of S-BPM in education and research.

2.1   Education

The degree course *Information Management* (IMA) as well as the same-titled post-graduate course both aim at holistic approaches to (re)-design and understand enterprises. As a consequence, we heavily focus on lectures covering business administration and strategy, modeling and re-engineering of business processes, defining underpinning business rules, and business activity monitoring concepts. Furthermore, we strongly address the design and construction of agile IT infrastructures dealing with effective transformation approaches of business processes into ERP systems and workflow runtimes using cutting-edge technology strictly following so-called service-oriented IT infrastructure (SOA) design blue-prints [32]. Therefore, lectures covering operating systems, computer networks, programming, database design, and digital media technology are likewise important in order to fulfill our mission statement: addressing the business-IT gap to create value with strong emphasis on customer orientation. Taken together, we particularly focus on teaching design patterns, procedure models, and tools based on well-known and globally accepted enterprise architecture design principles.

As S-BPM turned out to be very helpful in supporting our efforts generating a unique selling proposition for the respective fields of education, we introduced the procedure model of the subject-oriented business process modeling approach into several courses, namely "Messaging and Workflow Systems", "Interdisciplinary Project", and "Business Process Management". In the future, we will

---

[11] Even if it is not discussed more deeply in this paper, each BPM project is also a change project and therefore some issues have to be considered, e.g. maximum involvement of the affected people).

also introduce S-BPM in numerous lectures of our new master course Information Management.

Course Messaging and Workflow Systems. The subordinate objective of this course is to set up and administer a representative unified messaging infrastructure as well as a workflow management system. Special attention is paid to the design and functional implementation of simple business processes into workflow engines. Moreover, profound knowledge is delivered with respect to prominent BPM standards e.g., BPMN, WS-BPEL, XPDL in particular, and Web service technology in general.

In doing so, we commonly use Microsoft technology (e.g., BizTalk Server, Exchange Server, SharePoint Server) which is quite powerful from the technical point of view because of its modular system architecture, the extensibility, and rich functionality as well as its wide-spread acceptance on the market. From a business analyst's perspective, however, business process modeling and design tools from Microsoft are presently not intuitive enough to be efficiently utilized by these people. Thus, we sought for more user-friendly toolkits, narrowing the gap between process design on one hand and technical implementation into workflow engines on the other. Several business process modeling tools were evaluated. Among them, an Microsoft Visio plug-in as a BPMN 1.2 compliant editor and the jCOM1 BPM suite were studied in closer detail.

Students were asked to design and model our department's project request process using two different design principle paradigms, i.e. the activity centric "classical" approach of BPMN/WS-BPEL and the subject-oriented approach of jPASS!. The final goal was to fully automate both versions of the process under investigation meaning that both process definitions had to be operationally rolled out onto an appropriate runtime environment. As a final result, both approaches were discussed in detail.

The BPMN/WS-BPEL approach. Due to the fact that BPMN is not executable for itself, the BPMN-group established the following procedure model in order to functionally setup the process definition on Microsoft BizTalk Server:

1. model the process using the Microsoft Visio BPMN plug-in
2. convert the BPMN definition into WS-BPEL which is executable on numerous workflow runtimes
3. import the WS-BPEL process definition to BizTalk Server
4. execute the corresponding workflow

Since students in the fifth semester of study were not familiar with BPMN, the process under consideration was rather complex for them to handle. In a nutshell, five different roles dealing with approximately 80 activities and 25 decision gateways connected with corresponding sequence or message flows had to be modeled after the process was analyzed in detail from the stakeholders perspective. As mentioned above, the graphical representation of the project request process was designed with Process Modeler 5 for Microsoft Visio 2007[12]. This

---

[12] Download from http://www.itp-commerce.com

tool was chosen because of its inherent capability to transform BPMN processes into WS-BPEL, which is a XML-based business process execution language.

However, for a successful BPMN to WS-BPEL transformation, one-to-one mapping of corresponding elements is mandatory. But this is rather not the case. Numerous restrictions apply because WS-BPEL is exclusively block-oriented, following design rules of well-defined XML document specifications, whereas BPMN process definitions can be modeled almost arbitrarily. Consequently, two major routes along the path of the BPMN to WS-BPEL transformation process exist. The first one is to perform a BPMN process definition strictly following guidelines of the WS-BPEL design principle. This way, the content of the BPMN process definition might deviate a lot from the original business process being executed in the respective organization. The second choice is to code the WS-BPEL process by hand followed by a retrograde transformation of the WS-BPEL code to BPMN. This approach is quite dissatisfying because the resulting BPMN process description might also be far away from reality. Anyway, many restrictions apply, independently from the approach one decides to take, reflecting very much the general problem of the semantic mismatch between business and IT. Because of the on-site availability of Microsoft BizTalk Server 2006 (our department holds the status of a Microsoft IT Academy so that Microsoft software can be downloaded as needed) as well as the fact that this SOA integration tool hosts a powerful workflow runtime engine, the WS-BPEL XML code under consideration was rolled out to this process orchestration engine. Albeit the existence of a WS-BPEL import interface, the WS-BPEL process definition generated by means of Process Modeler 5 was not compatible with Microsoft BizTalk 2006. As a consequence, prior to importing, special adaption with respect to data types of numerous variables had to be performed by hand. Then, the import was successfully conducted.

Due to the fact that Microsoft BizTalk Server does not provide a client-side interaction interface out of the box, Microsoft Office InfoPath forms were to be created in order to allow participants of the business process to interact accordingly. Thus, in addition to accumulate knowledge about handling Microsoft BizTalk Server, students were quite challenged with respect to Microsoft Office InfoPath programming tasks. Taken together, it took them much time to satisfy all the requirements.

The S-BPM approach. The second group of students was asked to take an alternative pathway to model and deploy our department's project request process using the jCOM1 S-BPM suite as an intuitive means[13]. After they had become familiar with both, the innovative design paradigm of S-BPM as well as the related tool suite, students were able to effectively model the process under investigation. Quickly summarized, the S-BPM modeling procedure is as follows:

1. define all occurring roles of the respective process as so-called subjects

---

[13] We highly appreciate Dr. Albert Fleischmann's continuing interest in our work and thank him for providing the jCOM1 S-BPM suite for education and scientific project work.

2. set up the message flow between all subjects as needed
3. shape the internal behavior of each subject, i.e. build the individual subject's activities from its point of view (what is roughly comparable to a workflow within a lane of a BPMN process)
4. validate and simulate the process from the unique roles' perspective
5. roll out the process in the corresponding productive environment.

Steps 1 to 3 are performed using jPASS! whereas step 4 is carried out by jLIVE!. Finally, process execution is supported by jFLOW! (5).

Since the jCOM1 S-BPM suite represents all modules needed to successfully define executable workflows from scratch, no additional knowledge concerning other tools is needed along the complete path of the business process lifecycle. Moreover, no coding is needed when the associated workflow is executed within the intrinsic workflow runtime engine. Hence, this approach turns out to be very convenient for business analysts and, in our case, for students in particular, to fulfill even complex requirements. As a striking example, given the same requirements to both student groups, the BPMN/WS-BPEL approach mentioned above took about ten times longer being successfully delivered as compared to the S-BPM approach. This was mainly caused by the fact that the former approach is highly challenging with respect to exerting profound knowledge, touching various aspects of process and Web service standards (e.g. BPMN, WS-BPEL, SOAP, WSDL, XML) as well as programming, paired with extensive skills in the area of server technology (e.g., Microsoft BizTalk Server).

Contrarily, concerning flexibility and capability of interfacing, the BPMN/-WS-BPEL approach is much more flexible. This is due the fact that BPMN is a well accepted design standard for business processes in the business process analyst community and thence people are rather familiar with it. Furthermore, WS-BPEL is quite well supported by all prominent workflow orchestration engines. Although numerous derivates of WS-BPEL exist, this executable process language represents a common denominator thus being a powerful means to set up executable business processes in heterogeneous system environments.

It is worth mentioning that, using the jCOM1 S-BPM suite, it is also possible to generate WS-BPEL code because of the existence of a corresponding export interface. However, WS-BPEL can only be generated on a particular subject's base. As a consequence, the orchestration of the overall message flow between individual subjects has to be coded by hand after the individual subjects' WS-BPEL codes have been exported making the S-BPM approach likewise inconvenient, compared to the BPMN/WS-BPEL approach, rolling out process definitions in a heterogeneous environment.

Taken together, the BPMN/WS-BPEL approach turns out to be more flexible insofar as many tools exist supporting the transformation process starting from BPMN modeling and following the path over WS-BPEL to an appropriate target workflow runtime engine. In addition, due to the widespread distribution as well as acceptance of BPMN as an OMG standard, many business analysts and business process designers are very familiar with this notation. Since many enterprises, however, do not yet consider process design and programmatic process implementation as part of a holistic approach, both dimensions, i.e., business process management (BPM) and workflow management (WFM), will be performed independently from each other. Therefore, the phenomena of the so-called business-IT gap will never successfully converge.

In contrast, the S-BPM approach reveals an innovative and rather different approach with respect to BPM, as compared to the BPMN/WS-BPEL procedure model mentioned above. In this instance, the acting roles (e.g., persons, (IT) services or machines which reflect the subjects) are in the center of the design process, whereas the corresponding activities are modeled afterwards from each subject's perspective. This way reflects much more the real world situation concerning how people work and therefore, more mature and (cost-) effective business processes might result.

One big advantage of the jCOM1 S-BPM suite, in comparison to the BPMN/-WS-BPEL approach is that processes designed by means of jPASS! can be immediately validated by jLIVE! from the perspective of each role involved. As an important consequence, potential inconsistencies can be detected before the process under investigation is rolled out into the operative environment, thus avoiding problems before they might arise. A further big benefit of the S-BPM toolkit is that processes can be instantly embedded in the enterprise in an operative way where only a Web browser is needed as a fully featured client. Thus, end users commonly do not need any education in order to effectively take part in the process. One disadvantage of the S-BPM approach might be, however, that this toolset exhibits a monolithic architecture, thus preventing the process definitions being delivered to a heterogeneous system without transformation into WS-BPEL or any other executable process definition language. This might be a hindering reason for a broad implementation scope, e.g. business spanning business processes.

Course Interdisciplinary Projects

Business Rules and S-BPM. In one of our recent projects we were dealing with the integration of the business rules concept into S-BPM software. Business rules are instructions which are applied during the execution of a business process, linking back to a company's strategy. They need to be defined independently from the actual process and require a clearly defined business vocabulary. The approach was to derive the business vocabulary from XML Schema definitions which had already been used for the definition of messages in the S-BPM processes.

We used Microsoft BizTalk Server 2009 as business rules repository by uploading the XML Schema definitions to create a common business vocabulary as a basis to define business rules. That way, business rules, based on S-BPM processes could be defined within an external rule repository with the possibility to execute these rules with a dedicated rule engine. How to trigger the external execution of business rules from a process defined and executed in the jCom1 BPM Suite is one of our current research activities.

ERP Systems and S-BPM (application integration). In this project we wanted to check the connectivity of the jCOM1 BPM Suite. For this reason we designed a typical business process (e.g. purchasing) within the BPM Suite and investigated several possible communication scenarios with an "off the shelf" ERP system (Microsoft Dynamics NAV 2009) through web services. The basic idea was to define business processes within the best fitting application (typically some business process management and execution platform), but also to use existing assets such as ERP systems to gain a better integration of processes throughout the whole company. Beside technical aspects we also wanted to investigate the business aspects of such scenarios (economic value) [30].

Different scenarios have been investigated, but a communication between jCOM1 BPM Suite and Microsoft Dynamics NAV 2009 couldn't be established because of different technical issues, e.g. restricted handling of web-services on the side of jCOM1 and problems to call a NAV web service from java code (a problem of authentication) which could not be solved within the project schedule.

2.2 Research

Research efforts at our department are strictly aligned to the general strategic orientation of our university. As a university of applied sciences, we aim at research fields residing mainly in the area of applied research dimensions in order to stimulate spin-off effects to our local economic system. In particular, the research agenda at the department of Information Management deals very much with establishing holistic procedure models to describe enterprises from strategy to infrastructure design, independent from their respective branches with main emphasis on agility and adaptability. Due to the innovative design paradigm, S-BPM and dedicated tools attracted our special attention.

Two different but interrelated topics have been investigated so far in our laboratory. The first point of interest deals with the issue whether capability management and S-BPM can be merged into a unique procedure model to combine both, resource/asset and business process management, respectively, in an effective and sustainable way. The second topic of concern stresses the question if unified communications technology is a useful means to reflect human interaction behavior within an ad-hoc business processes ecosystem designed by S-BPM. The following chapters provide some insights into both topics and relevant results are being discussed in some detail.

Capability Management and S-BPM. Induced by an extremely challenging market, enterprises must gain the ability to adapt to changes of their environment in an effective as well as in an efficient way. This is why companies are forced to seek for new concepts and means which help them to become more adaptable and flexible in order to stay competitive.

Nowadays, business processes are commonly recognized to represent essential components of an enterprise architecture. Therefore, the management of business processes attracts special attention within a business ecosystem. Business processes, however, are commonly afflicted with high dynamics. These assets must be constantly adapted and/or changed in order to generate a sustainable increase in value. Since business processes describe how enterprises manufacture their products and services, their sequence of activities has to be adapted in case of changes, for example during integration of new partners or technologies as well as in case of changes in the enterprise strategy. Consequently, ongoing business performance is guaranteed and companies stay competitive and up-to-date.

The concept of the enterprise abilities or the so called business capabilities (BC) was deduced from the need to be able to handle the dynamics around and in enterprises successfully [31]:

A business capability is a particular ability or capacity that a business may possess or exchange to achieve a specific purpose or outcome. A capability describes what the business does (outcomes and service levels) that creates value for customers; for example, pay employee or ship product. A business capability abstracts and encapsulates the people, process/procedures, technology, and information into the essential building blocks needed to facilitate performance improvement and redesign analysis.

To our understanding, business capabilities represent abilities from the different areas of an enterprise to achieve a certain goal or result. They describe what an enterprise achieves in order to obtain customer value but they do not deal with resources needed to accomplish particular tasks. The definition of a capability contains no information about resources to be used and the sequence of activities or tasks in which they are applied. Hence, capabilities are rather stable and independent from change and dynamics in the enterprise whereas business processes as well as resources might change at a higher rate. To sum up, business capabilities represent a service-oriented-alike approach in the area of business agenda and can thus be seen as a counterpart to IT services.

Business capabilities simply define WHAT is to be done to achieve a certain goal. By allocation of resources such as services, technologies, information, and human resources, the question "WHAT IS DONE WITH" is answered. By assembling of appropriate capabilities to so called capability chains or by mapping of capabilities to process activities, information about the sequence flow emerges and the question HOW a certain product or service is constituted is answered. Figure 2 represents an overview of the interrelationship between business capabilities, resources, and business processes.

Taken together, due to an almost arbitrary coupling of stable capabilities with dynamic resources and processes as well as by a purposeful and highly flexible coupling of the business capabilities among themselves, a fast reorganization of services and assets should be facilitated, thus empowering an enterprise to react effectively as well as efficiently upon changing demands.

S-BPM and business capability management (BCM) seem to be highly innovative but orthogonal concepts. In course of our research efforts, we got interested in the possibility whether both concepts can be successfully combined or

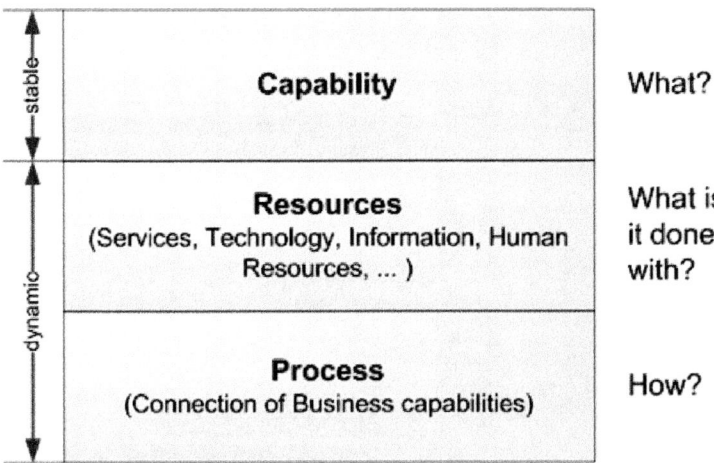

Fig. 2. Interrelationship between business capabilities, resources, and business processes)

not. Therefore, an appropriate research design was set up to prove the following hypothesis:

The transformation of business capabilities into business processes by means of the subject-oriented design paradigm is a valid as well as efficient path within the context of BPM.

In our approach, an already existing and well modeled business process (approval process for IT investments) was decomposed into business capabilities and a so called business capability map was established, based on a generically defined business capability map published by Sehmi and Schwegler [33]. A business capability map is a hierarchical and descriptive representation of business capabilities starting from top level foundation capabilities which are common to almost every enterprise despite of size and business (see Figure 3).

Starting from the foundation capability "Plan and Manage the Business", a high-level capability map was established (see Fig. 4). This high-level capability map was used as a scaffold in order to further extend the map, using already defined capabilities of the approval process under investigation. As a result, a comprehensive capability map at high granularity was constructed. Then, unique capabilities were put into logical relationship to obtain capability chains that represent particular task sequences which reflect process activities from the phenomenological point of view. This way, all parts of the respective business process were reassembled from corresponding capabilities so that a holistic sequence description by means of capability chains was available.

Afterwards, business capability chains were merged with S-BPM design components using jPASS!. As a first step, all subjects necessary to completely describe all roles of the process were defined and all dedicated messages between subjects were created. Next, the internal behavior of these subjects was modeled.

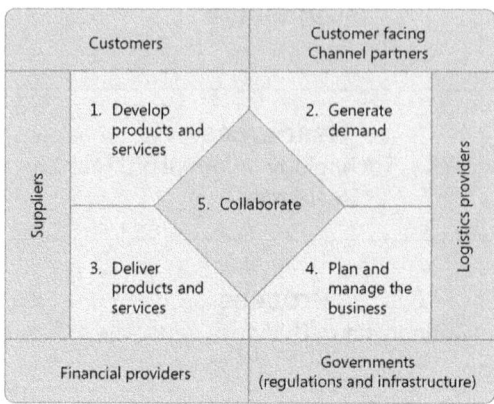

Fig. 3. Foundation capabilities of representative enterprises [33]

4.0 Plan and Manage the Business
  4.1. Financial Management
    4.1.5. Budgeting
      4.1.5.1. Manage Financial Resources
      4.1.5.1.2. Develop Budgets
  4.2 Project Management
    4.2.1 Project Management
      4.2.1.1 Create & Manage Projects
      4.1.1.2 Authorize Project Spending
      4.2.1.3 Manage Budget & Resources
      4.2.1.8 Manage Quality
      4.2.1.9 Monitor Status
    4.2.2 Project Accounting
      4.2.2.1 Set up Project Financials

Fig. 4. A high level capability map – "Plan and Manage the Business"

Therefore, the corresponding capability chains were transformed into appropriate functional states (e.g., function, send or receive state) and interconnecting state transitions were set up as required. As a consequence, the complete business process was successfully embedded within the jPASS! design environment with the assistance of the respective business capability, following the S-BPM design principle. Moreover, simulation and validation of the business process under investigation using the jLIVE! module was also successful.

Briefly summarized, the attempt to combine BCM and S-BPM turned out to be quite successful. Therefore, we were able to verify our working hypothesis that transformation of business capabilities into business processes by means of the subject-oriented design paradigm is a successful pathway.

Nevertheless, further investigations will be necessary with respect to the automation and tuning potential of the transformation procedure model, outlined above. Since the jCOM1 BPM suite currently has not yet the ability to access resources outside of the suite's cope in an efficient way, the development of adequate interfaces should be a major concern. Upon successful development of appropriate interaction layers, many different BPM and workflow orchestration engines, e.g. Microsoft BizTalk Server and Microsoft SharePoint services, IBM WebSphere, the inubit BPM-Suite or the TIBCO SOA product suite could be efficiently connected besides the already existing interoperability to the SAP business suite, thus leveraging the S-BPM design paradigm to a commonly accepted modeling principle not only for in-house but also for enterprise-spanning business processes.

2.3   Unified Communications and Human Interaction Management

Nowadays, real-time communication is a very important aspect within the context of an enterprise as well as in the context of exchange of information with suppliers and customers. Without an adequate communication strategy, firms perform poorly, thus running towards the risk of being eliminated from a hypercompetitive market. Not only cost-saving aspects turn out to be important drivers for implementing communication strategies and appropriate tools but also the human way of communication behavior seems to be an effective stimulus to introduce and manage human-interaction based business processes in enterprises. Since S-BPM highly reflects human interaction patterns, we became interested in how business processes and human interaction patterns need to be aligned to create sustainable business value.

The concept of Unified Communications (UC) combines different aspects under its hood. On the one hand, sociological and cognitive dimensions must be considered. On the other hand, various communication channels must be combined from a technological point of view, meaning that Audio and Video Conferencing (telephony, Web conferencing), Unified Messaging (Email, fax, voice mail, Short Message Service (SMS), Multimedia Messaging Service (MMS)), and Instant Messaging (Chat) have to be integrated in a reasonable way.

Numerous perceptions about UC exist. The following definition might reflect a representative compilation [34]:

Unified Communications is a new communication architecture, in which various forms of real-time communications and collaboration applications are integrated so individuals can manage all their communications together rather than separately, in both desktop and mobile environments.

This citation embraces the phenomenological appearance as well as the overall strategy of UC. It deals with an innovative communication architecture where numerous communication channels are holistically combined to become efficiently manageable and to provide pronounced usability. Furthermore, UC technology must be highly integrated in line-of-business applications (e.g., ERP, CRM or groupware clients, spreadsheet and text editors) covering both desktop and mobile workspaces. Finally, presence status information must be provided in order to reflect the availability of respective information and knowledge workers. In order to preserve privacy, the presence status must be under exclusive control of the employees.

In order to investigate how UC technology can be successfully integrated into ad-hoc business processes (i.e., human interaction workflows) we set up an appropriate research design in close cooperation with an IT systems integration company. Our working hypothesis was as follows:

Because S-BPM is highly message-oriented, it turns out to be a powerful approach to support human interaction business processes supported by underpinning UC technology.

At the very beginning of the project, a business process analysis was performed to identify eligible process candidates exhibiting qualified features to be supported by UC. We chose the pre-sales business process of *Datentechnik Austria GmbH & Co KG* which is highly driven by human interaction behavior and ad-hoc patterns. As a next step, the process was modeled with the aid of jPASS! and validation of the corresponding workflow was done by jLIVE!. Due to the fact that *Datentechnik Austria* wanted us to functionally integrate the pre-sales process into their existing CRM infrastructure, we did not make use of jFLOW! as workflow runtime environment. In addition, the jCOM1 BPM suite is currently not powerful enough to fully support UC components. This is, because appropriate interfaces to UC systems are not yet available.

The technical implementation of the process definition was realized by exclusive adoption of Microsoft technology. CRM and workflow orchestration features were provided through Microsoft Dynamics CRM 4.0 (i.e., CRM functions, workflow engine) whereas UC features were added through Microsoft Exchange Server 2007 and Microsoft Office Communications Server 2007 R2 (i.e. Email, Voice Mail, Instant Messaging, Telephony, Audio and Video Conferencing, Application Sharing). Presence status was enabled through directory services of Microsoft Active Directory. All desktop and mobile clients used were fully UC enabled.

Taken together, our experimental approach to combine S-BPM with UC technology was quite successful. We could clearly demonstrate that the S-BPM modeling paradigm is a powerful means to efficiently and effectively design human interaction business processes. To our interpretation, this can be mainly ascribed to the phenomenological analogy between the human interaction behavior, that is highly coined by message exchange and the manner how business processes are designed with jPASS!, were messages exchange is also designed between individual subjects as one of the first steps in the course of the business process modeling procedure. Furthermore, underpinning UC technology creates substantial value within the environment where human interaction plays a central role as communication pattern. In our case, the pre-sales process under investigation was boosted with respect to performance concerning the interaction between sales staff and customers. Relevant information was delivered to stakeholders in a very timely fashion, thus generating remarkably higher customer satisfaction which is very important during the initial business contact phase. Finally, upon introduction of the new UC architecture, high acceptance among employees was identified from the very beginning. This could mainly be attributed to the fact that humans feel very comfortable with systems that reflect their intrinsic way of working.

3   Conclusion and Future Work

Facing the challenge to become one of Austrias leading universities particularly in the field of applied sciences, we are highly interested in delivering continuous innovation to our students on one hand as well as to our customers on the other. Therefore, we are steadily seeking for exceptional approaches within the research and educational scopes of our department. In the course of these efforts, we persistently search for suitable partners from universities and industry to line up with us, thus guaranteeing mutual exchange of skills in order to meet the requirements defined as superordinate objectives of our university. Offering high-end equipment from the technological point of view corresponding with adequate expertise provided by highly-skilled staff, we are confidently encouraged being quite attractive for future students and for stakeholders from economy.

In this paper, we provided some insights into our strategy of complying with ever changing demands regarding research and education in the field of BPM. We impressively demonstrated that S-BPM was successfully introduced in numerous courses of our bachelor and master program, respectively, thus stimulating valuable discussions among our students concerning the past and present value of BPM as well as creating something like a new mindset regarding S-BPM. Moreover, we argue that S-BPM will stimulate pertinent research with respect to managerial and cost aspects of this innovative BPM paradigm. In addition, we are utterly convinced that new procedure models must be pursued in order to efficiently transform process definitions into IT infrastructures, where applicable, under the premise of flexibility and adaptability.

We believe that we are at a turning point of business process management. As a matter of fact, we have to assume that organizations have problems to implement business process management up to its full potential and we conclude from this fact that there is a need for improvement in certain areas.

Firstly, we have to develop more effective methods for the design and planning phase of the business process lifecycle. As a starting point, we propose further research based on empirical analysis to identify improvement opportunities based on facts and figures. Additionally existing literature on the use and implementation of business processes within organizations needs to be analyzed more stringently (meta studies).

Secondly, we need further industry-related research of how IT-systems could better support business process enactment and execution. We think special emphasis is needed on SMEs, as they are the backbone of most (even all) economic systems. Further research is needed to obliterate the business-IT-gap with innovative IT-systems. This means not only to develop theories and methodologies, but also to introduce innovative applications (i.e. software and systems) for business process management which overcome existing constraints. Business Process Management is not only the execution of strictly defined and fully automated processes through IT-systems, but also the management of human interaction. Therefore we need methodologies to handle communication and collaboration within a system, i.e. a firm.

Finally, we see a great need for education and training in this area in order to transfer actual research results into organizations. Therefore, new ways to teach business process management need to be developed and applied[14].

References

[1] Carr, N.G.: Does IT matter? Information technology and the corrosion of competitive advantage. Harvard Business School Press, Boston (2004)
[2] Hammer, M., Champy, J.: Reengineering The Corporation - A Manifesto for Business Revolution Harper Business (1993)
[3] Hammer, M.: The agenda: What every business must do to dominate the decade. Crown Business, New York (2001)
[4] Kosiol, E.: Organisation der Unternehmung. Gabler, Wies-baden (1962) (in German)
[5] Smith, H., Fingar, P.: IT Doesn't Matter – Business Processes Do. Meghan- Kiffer Press, Tampa (2003)
[6] Taylor, F.W.: The Principles of Scientific Management Harper & Bros., New York (1911)
[7] Smith, A.: Wealth of Nations Prometheus Books, New York (1991)
[8] Davenport, T.H.: Process Innovation – Reengineering Work through Information Technology. Harvard Business School Press, Boston (1993)
[9] Porter, M.E.: Competitive advantage: Creating and sustaining superior performance. Free Press, New York (1985)
[10] Weske, M.: Business Process Management. Springer, Berlin (2007)
[11] Dijkman, R.M., Dumas, M., Ouyang, C.: Semantics and analysis of business process models in BPMN. Information and Software Technology 50, 1281–1294 (2008)

---

[14] Actually an initiative, called SOUL, is under development; please contact one of the authors for more information

[12] Russell, N., ter Hofstede, A.H.M., van der Aalst, W.M.P., Mulyar, N.: Work-flow Control-Flow Patterns: A Revised Review, BPM Center Report BPM- 06-22 (2006), BPMcenter.org
[13] van der Aalst, W., van Hee, K.: Workflow Management. MIT Press, Cambridge (2004)
[14] van der Aalst, W.M.P., ter Hofstede, A.H.M.: YAWL: Yet Another Workflow Language. Information Systems 30(4), 245–275 (2005)
[15] ISO/IEC 19502:2005 Information technology – Meta Object Facility (MOF)
[16] Schmidt, W., Fleischmann, A., Gilbert, O.: Subject-Oriented Business Process Management. HMD – Praxis der Wirtschaftsinformatik 266, 52–62 (2009)
[17] Smith, H., Fingar, P.: Business Process Management – the third wave. Meghan-Kiffer Press, Tampa (2007)
[18] Smith, H.: Business process management the third wave: business process mod-elling language (bpml) and its pi-calculus foundations. Information and Software Technology 45, 1065–1069 (2003)
[19] Smith, H., Fingar, P.: Workflow is just a Pi process, White Paper (January 2002), http://www.bptrends.com
[20] http://de.wikipedia.org/wiki/Business_Process_Modeling_Language (viewed on 01.05.2010)
[21] Zilio, D.S.: Mobile Processes: a Commented Bibliography
[22] Milner, R.: Calculus of Communicating Systems. Springer, Berlin (1998)
[23] Milner, R., Parrow, J., Walker, D.: A calculus of mobile processes, Part I/II. Information and Computation 100, 1–77 (1992)
[24] Hoare, C.A.R.: Communicating Sequencial Processes. Prentice Hall, New Jersey (1985), http://www.usingcsp.com/ (check on 05.01.2010)
[25] Puhlmann, F.: Why Do We Actually Need the Pi-Calculus for Business Process Management? In: Abramowicz, W., et al. (eds.) Proceedings of the 9th Interna-tional Conference on Business Information Systems (BIS 2006). LNI, vol. P-85, pp. 77–89 (2006)
[26] Puhlmann, F., Weske, M.: A Look Around the Corner: The Pi-Calculus. In: Jensen, K., van der Aalst, W.M.P. (eds.) Transactions on Petri Nets. LNCS, vol. 5460, pp. 64–78. Springer, Heidelberg (2009)
[27] van der Aalst, W.M.P.: Why Workow is NOT just a Pi process, White Paper (February 2002), http://www.bptrends.com
[28] Fleischmann, A.: Distributed Systems – Software design and Implementations. Springer, Berlin (1994)
[29] Reinertsen, D.G.: Managing the Design Factory. Free Press, New York (1997)
[30] Singer, R.: Integration of Microsoft Dynamics NAV 2009 with Process Manage-ment Suite jCOM1. In: Microsoft Convergence 2009, Vienna, October 28 (2009)
[31] Homann, U.: A Business-Oriented Foundation for Service Orientation, Microsoft Corporation (February 2006), http://msdn.microsoft.com/en-us/library/aa479368.aspx (accessed on February 3, 2010)
[32] Ortner, W., Tschandl, M., Zinser, E.: Potentials of Service-Oriented Architec-tures. In: Erkollar, A. (ed.) Enterprise & Business Management – A Handbook for Educators, Consulters and Practitioners, pp. 226–267. Tectum Verlag, Mar-burg (2007)

[33] Sehmi, A., Schwegler, B.: Service-Oriented Modeling for Connected Systems - Part 1, Microsoft Corporation. The Architecture Journal 7, 33–41 (2006)

[34] Lazar, I.: Unified Communications: What, Why and How? What Value Can Unified Communications Bring to Your Enterprise (2007), http://whitepapers.techrepublic.com.com/abstract.aspx?docid=335624 (accessed February 4, 2010)

Application of Subject-Oriented Modeling in Automatic Service Composition

Erwin Aitenbichler and Stephan Borgert

Technische Universität Darmstadt, Hochschulstrasse 10, 64289 Darmstadt, Germany

Abstract. Next generation SOA systems promise to enable an "Internet of Services" (IoS) - an open environment, in which every participant is free to offer and consume services. Such an IoS gives businesses the opportunity to outsource parts of their internal processes and to replace them by using external services. However, businesses must ensure that external services are compatible with their processes and that they can quickly adapt if service offering changes on the market. This raises the need for a process definition language with a formal foundation and well-defined semantics. In this paper, we discuss the suitability of different process definition languages for automatic service composition, show that subject-oriented modeling with PASS is well-suited for this domain, and how automatic service composition is implemented in the Theseus/TEXO project.

1 Introduction

Service-oriented Architecture (SOA) is an architectural style that facilitates loose coupling of components, and consequently enables flexible selection and substitution of services. However, todays SOA systems are rather closed. They are only used within the boundaries of an enterprise, or sometimes within conglomerates of enterprises with long-standing cooperations. To match the reality of Business Value Networks, current systems must evolve towards open service environments.

A *Business Value Network* (BVN) emerges from dynamic interactions of loosely-coupled organizations, which are legally distinct but economically interdependent, performing different value-creating roles (e.g., suppliers, distributors, service providers, infrastructure providers) that leverage their core competencies in order to flexibly craft optimum response to rapidly changing markets and customer demands. Value is created via dynamic exchanges of shared information and resources among these organizations engaged in complex and co-evolving processes wherein dominant players can shape the network context [1].

The term *Future Business Value Network* (FBVN) inherits this concept and stands for a conceptional framework which describes organization models with configurations of value adding collaborations within cooperative social networks among enterprises, (public) organizations, and individuals. A further characteristic is the aim to achieve a common set of goals enabled through the *Internet of Services* (or any other upcoming technology framework). FBVNs are motivated by the marching processes of outsourcing, tertiarisation, globalization, and technical innovation.

The basis for such an Internet of Services is currently developed in the large-scale Theseus Programme [1]. Building on the notion of a SOA, interacting software components can be loosely coupled and distributed over the Internet. The Theseus/TEXO

H. Buchwald et al. (Eds.): S-BPM ONE, CCIS 85, pp. 71–82, 2010.

platform allows for a fully decentralized service provisioning, since service consumers and providers are communicating directly with each other, in a peer-to-peer manner. The market participants are brought together by a number of central entities, such as the service marketplace and the community portal.

Such an Internet of Services gives businesses the opportunity to outsource parts of their internal processes and to replace them by using external services. In an open service market, where anybody is allowed to offer services, it seems natural that there will be numerous offers for services providing the same functionality. Hence, the customer can leverage the effects of an open market, concentrate on his core business, and save costs. However, the services offered will still be different in many details, such as their quality and how their internal processes are realized. Consequently, the customer in a B2B scenario must ensure that his overall process and the processes of external services are compatible with each other.

For example, public institutions in Germany have to stick to a clearly defined buying process. It defines how the institution has to verify the past behavior of a supplier and that he has not been blacklisted, how offers have to be invited, how offer evaluation meetings have to be organized, how offers have to be evaluated, which order approvals are necessary, how orders have to be made, and how payment is made. It also defines that suppliers cannot ask for pre-payment and must not charge shipping costs. Now, if a supplier would insist on pre-payment, the buying process would fail, because the customer is not allowed to do so.

However, the main issue with this example is the time when the process incompatibility is discovered: in the middle of the process - which is much too late. Similarly, process compatibility is an important aspect in the automotive industry. The processes for ordering components at suppliers, shipping to the car manufacturer, payment, etc. must match and all activities must be executed in the correct sequence and at the right time, such that the overall process is successful and completes in the designated time.

Consequently, one important aspect is to verify - before a service is purchased and the process is executed - that all potential messages flowing between the process participants can be handled adequately, that all activities are executed in the correct sequence, and that the process eventually terminates. In order to test process compatibility automatically, this first raises the need for a suitable process description language. Because the intricacies of how process models are described and maintained are rooted in Business Process Management (BPM), we first discuss the current state of BPM in industry and the associated mainstream process description techniques in Section 2. Next, we describe Subject-oriented BPM and the PASS language, which are the basis of our automatic service composition approach presented in Section 3. In Section 4, we present the current state of the automatic service composition implementation in Theseus/TEXO. Related work is discussed in Section 5. Finally, the paper is concluded in Section 6.

2   Current Issues in BPM

The mainstream process description languages used today lack a formal foundation and well-defined semantics (e.g., EPC, BPMN), or they are too low-level (e.g., BPEL). Consequently, such description formats do not permit computers to reason about processes. Beside these technical aspects, BPM also suffers from several other problems, as current practice in the implementation of BPM projects shows.

2.1 Lack of Process Governance

A first fundamental problem is that processes are not 'lived' as they have been designed and modeled. Practitioners report that the vast majority of decisions made in a business are still based on gut feel, intuition and experience. "We think the process works like this, so we should do X?" or "Customer orders were delayed in the past primarily because of Y, so go fix that!" [2].

If processes had initially been modeled, then the corresponding models are often not kept up-to-date. A recent study by Gartner reveals that many BPM projects will fail after implementation because the proper supporting disciplines are not implemented: "Too many user organizations are adopting BPM technologies without applying BPM disciplines via the competency center, and find that their efforts do not deliver the promised results, and their BPM initiatives are disbanded." [3]. Similarly, Forrester underlines that process support is not only about technology: "Too many organizations believe they can implement BPM with nothing more than a comprehensive set of tools and a good return on investment story" [4].

The lack of up-to-date process models also impedes the assurance of process quality, analysis of process efficiency, and process improvement. Alone the discovery of how a process actually works in a business can cause significant costs: BPM consultants claim that they spend around 40% of the project time finding out how processes in a business actually run.

2.2 One-Shot Transformations

When parts of a process are implemented as software components, such as Web Services, then often so-called one-shot transformations are made. For example, the implementation starts with a business analyst creating a BPMN model of the process, which emphasizes the business aspects. However, from a technical point of view, these models are abstract, inexact, and omit many essential technical details that would be necessary to be able to execute the process directly on a computer. Next, a software engineer creates an executable BPEL process model based on the BPMN. This is either done manually or by means of automatic transformations. However, because BPMN lacks formal semantics, such a transformation can at best produce a "BPEL skeleton", which contains the structure of the process, but the engineer has to fill in all the technical details manually. Consequently, engineers transform from abstract to more concrete models and add details to the model. The relationships between model elements from the concrete to the abstract model get lost and it is not possible to automatically update the abstract model containing the business perspective.

2.3 "Outlook Processes"

Another common implementation of processes are so-called "Outlook processes". Such processes contain activities like "send email to financial accounting", meaning that an employee uses ordinary email to perform a process step. Such processes have two major drawbacks.

First, the progress of process instances cannot be monitored directly. If a customer asks for the current state of a process instance, it boils down to locating the person who had last acted on the process and asking her. While mail server products, such as MS Exchange, support message tracking, the relationships between mails and process instances cannot be discovered easily.

Second, because the email client allows unbounded communication, new communication paths between process participants can emerge easily. In reality, this changes the process, but this change will usually not make it back into the process model.

3   Process Modeling

In the following section, we discuss the requirements for a process modeling language as needed for our automatic service composition approach.

3.1   Requirements

Some important aspects of this process modeling language are:

- The process description language needs **formal semantics**. This property is needed to test the compatibility of processes, e.g., of the main process with the subprocess implemented by an outsourcing partner and to verify with formal methods that the process is correct.
- The description must have a **subject-oriented perspective**. Beside the description of *what* is done in the process, it must be clearly stated for each activity, *who* is responsible for it. This is important to decompose the overall process and to identify its constituent subjects for which services are inserted.
- The process model must be **directly executable** or it must be transformable to an executable process format without the need to manually add details to the generated model.
- The model must be **hierarchable**, i.e., it must be possible to move up and down in terms of the abstraction level. It should enable arbitrary refining or clustering of behavior without the need to leave the model. This is the key feature to eliminate one-shot model transformations.
- The language should equally support **software services** as well as **human services**.

To enforce governance, a valid process model must be in the information system that corresponds to the execution of the process in real world at all times. This is not only a requirement for automatic service composition, it is rather a general requirement to drive the next generation of BPM.

Instead of "Outlook Processes", execution platforms for the business process are needed, which do not allow interactions between participants beside the process. This forces the business to stick to the modeled process. There are no costs for discovering how the process works when it should be evaluated or made more efficient. Of course, this comes with a cost. For example, the business has to install several key users that are responsible for maintaining the process model. This approach has been successfully shown in [5].

Given these requirements, we have chosen subject-oriented modeling and the description language PASS.

3.2   Subject-Oriented Modeling

Subject-orientation introduces an approach that gives balanced consideration to the actors in business processes (persons and systems as subjects), their actions (predicates), and their goals or the subject matter of their actions (objects) [6,7]. It is based on the fact that humans, machines, and software services can be modeled in the same manner. Every one of them receives and delivers information by exchanging messages. Humans, e.g., exchanges emails, office documents, or voice messages.

3.3   PASS

The *Parallel Activities Specification Scheme* (PASS) [8] language is an implementation of subject-oriented modeling. It is founded on top of the process algebra CCS [9] (Calculus of Communicating Systems) and all language constructs of PASS can be transformed down to pure CCS. Process algebras provide a suitable means for modeling distributed systems. They offer well-studied algorithms for verification and for determining behavioral equivalences. In addition, the CCS composition operator facilitates a hierarchization and modularization of the model, allowing to handle business processes of arbitrary size. At the basic level, PASS only distinguishes between three basic types of activities: *send message*, *receive message*, and *function*.

3.4   PASS Extensions

To describe process patterns, the PASS language was extended. In contrast to regular PASS graphs, process patterns do not have to be fully connected graphs and may contain wildcard operators. Process patterns are used for service matching and their modeling differs from that of fully-specified processes in the following two aspects:

- In the model of a pattern, only activities are specified, which are essential for the process. This simplifies modeling, because the service engineer does not have to specify all functionalities and does not have to take care about each detail activity. E.g., he could omit modeling the payment branch of the process (, because its details might not be vital from a customer's point of view). If services have such branches, they would still be included, unless the engineer explicitly models the exclusion of certain behavior.
- The order of activities can be defined in a more general way as in usual process models. The wildcard operator can be used in conjunction with multiple isolated subgraphs to express a logical order between activities, instead of a single sequential order. This is useful, e.g., to enforce a certain behavior or communication pattern, while only concentrating on the essential parts of a process.

4   Automatic Service Composition

The desired result of service composition is specified by the *composition goal*. The goal consists of a description of the overall behavior, functional, and non-functional properties. In the following, we concentrate on the description of the behavior.

The behavior of the composition is described by a *fragmented* PASS process model. This model is underspecified, i.e., it only specifies the essential and basic parts of the desired process, but omits unimportant details. Consequently, it contains all subjects participating in the process and for each subject, it may contain a *process pattern* instead of a fully-specified process. Process patterns are used later to search for suitable services. We denote a model as being *fully specified*, if the behavior of all its subjects is fully specified. Models containing one or more process patterns are denoted as *fragmented models*.

Fragmented models should neither be overspecified nor be underspecified. If a pattern is overspecified, then the likelihood to find suitable services implementing this process diminishes. On the other hand, if a process is underspecified, then service candidates may bring unwanted behavior into the composition.

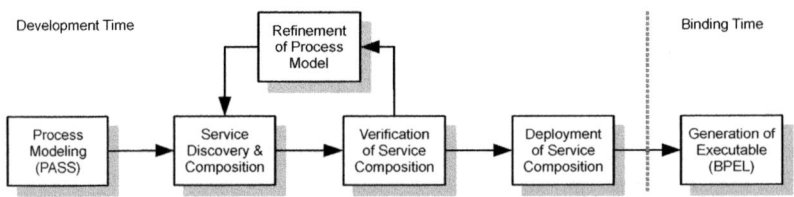

Fig. 1. Development process from fragmented model to executable process.

To specify the composition goal, we start with modeling an underspecified process and then refine it as far as needed. These steps are supported by tools. The development process is shown in Figure 1 and involves the following steps:

1. **Modeling:** In the first step, the initial fragmented PASS model is created. This model is typically underspecified.
2. **Discovery and Composition:** All suitable services are discovered according to the process patterns and constraints specified by the functional and non-functional properties. Then, a list of possible compositions is constructed.
3. **Verification:** While the compositions constructed in the previous step already match the structure of the desired process and its constraints, some compositions might still violate formal properties. In this step, each composition is entirely transformed into a CCS expression and then verified with formal methods.
4. **Refinement:** The developer inspects the service compositions found by the system in the refinement editor. Because the model is initially underspecified, the system might pull in services that expose unwanted behavior. The developer can exclude such behavior by refining the fragmented model.
5. **Deployment:** Once the fragmented model has been sufficiently refined, it can be deployed on the automatic service composition server. The server generates an executable BPEL process based on the fragmented PASS model and the candidate services. The server can also periodically repeat the discovery, composition, verification, and generation steps to take new services that appear on the market into account.

4.1   Modeling

The process is modeled using the Eclipse-based editor jPASS by jCOM1 [10]. Figure 2 shows the Subject Interaction Diagram of the TEXO EcoCalculator demonstrator.

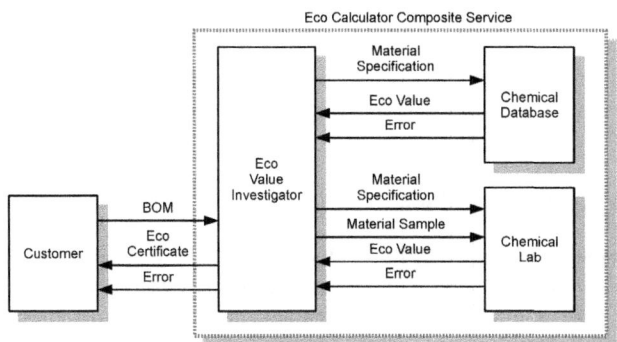

Fig. 2. Subject Interaction Diagram of EcoCalculator

This diagram describes the relationships between subjects and the types of messages exchanged. In this scenario, a government agency establishes a new "eco label" for cars meeting certain ecological requirements. One of the requirements is a concept for disassembling and recycling and the restricted use of certain environmentally harmful materials. The service provides a compliance check and cost simulation. Its process involves the following subjects:

- **Customer:** An OEM can invoke the EcoCalculator service by sending a Bill Of Material (BOM) to the service.
- **Investigator:** The Investigator is the main part of the EcoCalculator composite service. It implements the government policy and orchestrates additional services. The functionalities Chemical Database and Chemical Lab are provided by external services.
- **Chemical Database:** Third party service that provides detailed chemical information about materials.
- **Chemical Lab:** Third party service that provides a chemical analysis of (physically) provided material samples.

Figure 3 shows the fragmented process model of the subject Chemical Database. A fragmented model only describes the basic and essential parts of a process. Later, during

Fig. 3. Fragmented process model of the subject Chemical Database

service discovery, it is used as a search pattern to identify matching candidate services. In contrast to fully specified process, the fragmented model may contain *wildcard activities*. Such a wildcard matches any sequence of activities in the fully specified model of a service.

To make the matching of activities work, it is also vitally important that the activities in the search pattern and in the process description of a service are modeled using the same vocabulary. To ensure a consistent modeling, we introduce a so-called Activity Catalogue. It is a taxonomy of possible service functions and is based on the NACE catalogue [11] which is a statistical classification of economic activities in the European Community.

4.2 Discovery and Composition

The first step in composition is to find matching service candidates. To match the fragmented process with service descriptions, we use the programmed graph rewriting system GRL [12]. GRL stands for Graph Rewrite Library and is a Java library that provides the core functions of a graph rewriting system by supporting queries and rewrite operations. Rewrite rules are described in the declarative language GRL-RDL (Rule Description Language). GRL operates on directed, attributed graphs, whose data structures are defined by the respective application. Nodes and edges of the graph can be attributed by arbitrary Java objects. Its basic building blocks are predicates (tests) and productions (rewrite rules). Rewrite rules are specified textually. Complex attribute tests and transformations can be performed by calling Java methods from inside RDL programs. RDL programs are compiled, optimized using a heuristic, and then executed on a virtual machine. Hence, GRL provides highly efficient graph matching.

The service descriptions are used as work graphs and the goal specifications are translated into query expressions in the language GRL-RDL. To match the pattern with services, it is required that each service comes with a fully-specified PASS description. Applying graph algorithms leads to candidate lists for each specified pattern. The fragments defined in the first step are used to discover candidate services.

4.3 Verification

In order to verify the correctness of a possible service composition, the first test is to check the statical interfaces between the services. This involves the comparison of the message types exchanged between the respective services. Next, the dynamic interface is checked. This represents the communication behavior during runtime, such as the order of messages that are exchanged.

While the graphical representation is suitable for matching process patterns with service candidates, it is not suitable for verification. Hence, the PASS graphs are transformed to a pure CCS description, which is then used as input for the verification algorithms.

We currently use the CWB-NC Workbench [13] for running the verifications. CWB-NC supports various behavioral equivalences as well as model checks. Firstly, this allows us to identify services that expose equivalent behavior. At runtime, such services might be used as a replacement, in case that the original service fails. Secondly, a choreography conformance check can be performed. In a valid composition, it must be ensured that the involved services are able to communicate with each other.

4.4   Refinement

After the verification step, there may still be unwanted compositions, because the fragmented model of the service composition is initially underspecified. For each possible composition, the system now generates a fully-specified process graph by combining the process descriptions of its constituent services. The resulting process graphs are displayed in the refinement editor, where the service engineer can now annotate activities and eliminate unwanted behavior. An activity can have one of the following four states in the refinement editor:

- **required:** The activity is required and must be part of the resulting process.
- **forbidden:** The activity must not be part of the resulting process.
- **allowed:** The activity can be tolerated. It is not considered to be an essential functionality of the process.
- **unclassified:** The activity does not carry an annotation. This is the initial state of each activity.

4.5   Deployment

To determine all possible combinations of services, the first step was to discover all candidate services using process graph pattern matching. Next, these combinations were checked in the verification step and all incorrect combinations were discarded. Finally, for each valid combination, an executable BPEL process is generated, which orchestrates the constituent services.

5   Related Work

Several recent research efforts have focused on dynamic service composition techniques. Most of them are working on the execution level and extend the functionality of the BPEL standard by using proxy services or additional annotations or descriptions. A representative work is described in [14], where the authors introduce the VxBPEL language, which is an extension of BPEL by incorporating variability. It enables rebinding of services during runtime, substitution of service for optimizing purposes or in case of in sudden unavailable services. In contrast to our approach, choreographies are not supported. In addition, services provided by humans are not considered and there is a shortage of formal verification techniques.

A Petri Net is a formal language for modeling concurrent systems and has been widely accepted as formal foundation for business process modeling. Furthermore, it provides a graphical and easily understandable notation. Petri Nets are object of research for many years and current efforts are focusing on suitable constructs for automatic composition and choreography descriptions. For example, Huangfu et. al. [15] present an approach that addresses the issue of dynamic service composition by modeling service behavior by Object Petri Nets. A service consists of a set of operations and the paper introduces mapping rules from services to Object Petri Nets. A main drawback of using Petri Nets is that the entire process has to modeled in a single net. In contrast to this, we use a process algebra that supports parallelism. That allows to model each service separately and then compose them simply by using the parallel operator.

Process algebras like π-calculus provide strong means for modeling concurrent systems like service compositions and are based on formal terms. Choreography modeling, refining, and clustering are inherently supported. In addition, a rich theory to analyze processes for equivalence is provided and also the capability to perform reasoning on system properties and to verify process behavior. For this reason, current research efforts in this area focus mainly on approaches for formal verification of services and business process. Work on compliance and compatibility checks investigate the issue of when a service can be replaced by another one [16,17]. This is necessary when a service of a process fails during runtime or for finding redundant services. COWS [18] and SOCK [19] are designed for the purpose of automatic service composition. Furthermore, process algebras are often combined with other formalisms in order to be able to specify more aspects of a service in a formal manner. E.g., some extensions exists, that combine the π-calculus with ontologies [20,21] and formal logics [20,21] to describe non-functional properties and access control policies. While these formal approaches are also capable of formal verification and matchmaking, they usually do not consider other aspects, such as the execution of the models, or the seamless integration of human services. In contrast to Petri Nets, process algebras lack a graphical representation.

Human-centered process modeling is another area of growing research interest. Previous approaches for automatic service composition take mostly only software-based services into account. Work on human-centered process modeling is very technology-oriented and lacks formal methods for verification [22,23,24]. Also, choreography is not supported, since most approaches are based on extending BPEL.

6   Conclusion

In this paper, we have presented a novel approach for automatic service composition, based on process pattern matching.

An important prerequisite for this approach is a suitable process description language. While business analysts are very comfortable with visualizing business processes in a flow-chart format, most such formats used today in industry can only establish an interoperability on the human level. This creates a technical gap between the format of the initial design of the business process and the formats for verification and execution. In contrast to this, the PASS language fulfills three important properties:

- The formal foundation based on CCS allows formal verifiability.
- Its well-defined semantics allows direct execution.
- Its graphical representation is easily comprehensible by humans.

We have extended the PASS language with constructs to describe fragmented processes. This allows an engineer to describe the goal of a service composition, while focusing only on the essential and basic aspects of the process. In addition, we have presented a method for automatic service composition based on matching such process descriptions.

Acknowledgments. This work was supported by the Theseus Programme, funded by the German Federal Ministry of Economy and Technology under the promotional reference 01MQ07012.

References

1. BMWi: TEXO – Business Webs in the Internet of Services (2009),
   http://theseus-programm.de/scenarios/en/texo.html
2. Lees, M.: BPM Done Right: 15 Ways To Succeed Where Others Have Failed. Software AG (March 2008)
3. Olding, E., Cantara, M.: Highlights from BPM Summit. Gartner, Inc., London (March 2009)
4. Savvas, A.: Cultural Resistance Main Cause of BPM Project Failure. Computer Weekly (March 2005)
5. Konjack, G., Heckmaier, M.: AST – Order Control Process. In: Buchwald, H., et al. (eds.) S-BPM ONE. CCIS, vol. 85, pp. 117–122. Springer, Heidelberg (2010)
6. Fleischmann, A., Lippe, S., Meyer, N., Stary, C.: Coherent Task Modeling and Execution Based on Subject-Oriented Representations. In: England, D., Palanque, P., Vanderdonckt, J., Wild, P.J. (eds.) Task Models and Diagrams for User Interface Design. LNCS, vol. 5963, pp. 78–91. Springer, Heidelberg (2010)
7. Schmidt, W., Fleischmann, A., Gilbert, O.: Subjektorientiertes Geschäftsprozessmanagement. HMD - Praxis der Wirtschaftsinformatik (266) (April 2009)
8. Fleischmann, A.: Distributed Systems: Software Design and Implementation. Springer, Heidelberg (1994)
9. Milner, R. (ed.): Communication and Concurrency. Prentice Hall PTR, Englewood Cliffs (1995)
10. jCOM1: Welcome to the Future of BPM: S-BPM (2010), http://www.jcom1.com
11. NACE: Revision 2 (2010),
    http://ec.europa.eu/eurostat/ramon/nomenclatures/index.cfm?
    TargetUrl=LST_NOM_DTL&StrNom=NACE_REV2
12. Aitenbichler, E.: Entwurf und Implementierung eines programmierten Graphersetzungssystems in Java. Master's thesis, Institut für Technische Informatik und Telematik, Johannes Kepler Universität Linz (2000)
13. CWB-NC: The Concurrency Workbench of the New Century (2000), http://www.cs.sunysb.edu/~cwb/
14. Koning, M., Sun, C., Sinnema, M., Avgeriou, P.: VxBPEL: Supporting Variability for Web Services in BPEL. Information and Software Technology 51(2), 258–269 (2009)
15. Huangfu, X., Shu, Z., Chen, H., Luo, X.: Research on Dynamic Service Composition Based on Object Petri Net for the Networked Information System. In: Fifth International Joint Conference on INC, IMS and IDC, pp. 1075–1080 (2009)
16. Wu, Z., Deng, S., Li, Y., Wu, J.: Computing Compatibility in Dynamic Service Composition. Knowledge and Information Systems 19(1), 107–129 (2008)
17. Bordeaux, L., Salaun, S., Berardi, D., Mecella, M.: When are Two Web Services Compatible. In: Shan, M.-C., Dayal, U., Hsu, M. (eds.) TES 2004. LNCS, vol. 3324, pp. 15–28. Springer, Heidelberg (2005)
18. Lapadula, A., Pugliese, R., Tiezzi, F.: A Calculus for Orchestration of Web Services. In: De Nicola, R. (ed.) ESOP 2007. LNCS, vol. 4421, pp. 33–47. Springer, Heidelberg (2007)
19. Guidi, C., Lucchi, R., Gorrieri, R., Busi, N., Zavattaro, G.: SOCK: A Calculus for Service Oriented Computing. In: Dan, A., Lamersdorf, W. (eds.) ICSOC 2006. LNCS, vol. 4294, pp. 327–338. Springer, Heidelberg (2006)
20. Agarwal, S., Rudolph, S., Abecker, A.: Semantic Description of Distributed Business Processes. In: Proceedings of AAAI Spring Symposium – AI Meets Business Rules and Process Management (2008)
21. Markovic, I., Pereira, A.C., Stojanovic, N.: A Framework for Querying in Business Process Modelling. In: Multikonferenz Wirtschaftsinformatik, pp. 1703–1714 (2008)

22. Canfora, G., Penta, M.D., Lombardi, P., Villani, M.L.: Dynamic Composition of Web Applications in Human-Centered Processes. In: Proceedings of the ICSE Workshop on Principles of Engineering Service Oriented Systems, pp. 50–57 (2009)
23. Schall, D., Truong, H.L., Dustdar, S.: Unifying Human and Software Services in Web-Scale Collaborations. IEEE Internet Computing 12(3), 62–68 (2008)
24. Soriano, J., Lizcano, D., Hierro, J.J., Reyes, M., Schroth, C., Janner, T.: Enhancing User-Service Interaction through a Global User-Centric Approach to SOA. In: Fourth International Conference on Networking and Services (icns 2008), pp. 194–203 (2008)

Part II

Essential Capabilities

What Is S-BPM?

Albert Fleischmann

JCOM1, Lilienthalstraße 17,
85296 Rohrbach, Germany
Albert.Fleischmann@jcom1.com

Abstract. Subject-oriented Business Process Management (S-BPM) has emerged to a semantic paradigm, modeling, and implementation approach. This contribution reflects its current state of development. The review of underlying concepts and resulting benefits demonstrates the orientation towards actor responsibilities and communication transparency. The introduction of advanced modeling features reflects the capability of S-BPM to capture complex business cases while ensuring operational coherence.

Keywords: natural language, subject orientation, epistemological analysis, granularity, specificity, modeling constructs.

1 Introduction

In our global world, the division of work is increasing. This means coordination effort also increases in order to produce the required products or services. The business world has become more complex. Customers, suppliers and all the other internal or external stakeholders of an organization have to interact with each other as part of their day-to-day business. The parties involved in producing products or services have to agree on interaction behaviors for synchronizing their activities. This includes sequence and time constraints as well as other alternatives. Business processes define at what point in time process participants execute individual activities on related objects. During recent years, the focus in Business Process Management (BPM) was on modeling. In this case the emphasis was placed on analyzing and optimizing the resulting models. These optimized business models have traditionally been used for executing business processes. Because of rapidly changing business requirements, the corresponding business processes have to be adopted with the same speed.

A new generation of BPM must meet all of these requirements. In order to achieve this goal BPM is combined with SOA and Web 2.0. In [2] this approach to BPM is termed BPM 2.0. In [3] BPM 2.0 is considered less technology-oriented compared to the original concept. Herein, BPM 2.0 is also considered in conjunction with Enterprise 2.0, an approach for self-structuring organizations.

Our approach merges all of the various properties of BPM 2.0. BPM 2.0 has not only technical, but also human aspects. We have developed a new BPM methodology which combines all of these various properties of BPM 2.0, from a technical perspective as well as with regard to human interaction. We term this new innovative

H. Buchwald et al. (Eds.): S-BPM ONE, CCIS 85, pp. 85–106, 2010.

approach S-BPM. In S-BPM the focus is put on the acting elements within a process, the so-called subjects. Subjects execute and synchronize their activities by exchanging messages - a simple approach, based on the structure of sentences (subject, predicate and object) in natural languages.

In a complete active sentence the subject is the initiator of an activity, the activity itself is the predicate and the target of the activity the object of the sentence. In the following sections, we will show how S-BPM meets the requirements of BPM 2.0 and how the various computer science concepts are used to define this approach.

2 Properties of BPM 2.0

With BPM 2.0 it should be possible to react rapidly to changing business environments in a complex business world. In order to reach this goal a BPM 2.0 approach must have the following properties [1][2][3]:

First of all, the process users should be able to build and adapt process models by themselves. There should be no necessity for process modeling specialists or process implementation specialists. Only the participants in a process truly understand the complexity of the processes they are involved in.

Secondly, the models should be executable without any additional programming or programming know-how since process users do not generally have this knowledge. "Modeling by itself is not BPM. BPM 2.0 requires an integrated design and runtime environment – a BPM suite – that automates, integrates, and monitors process execution end-to-end."[1].

Thirdly, the process environment - the socio-technical system consisting of people, machines and software - should be easily integrated with the BPM model.

And finally, process execution should be measureable without a huge amount of additional effort since, at the end of the day, you want to know what advantages (and drawbacks) a process has.

3 Aspects of BPM 2.0

If we want to identify a BPM 2.0 approach which meets the requirements above, we have to consider various aspects.

We have to make an assumption about human nature. In Taylorism we assume that there are people who work and others who define how these people have to work. In a fast-changing world in which more and more people are highly educated, this approach is no longer adequate. In BPM 2.0 we want people to take on the responsibility for defining their own processes. This requires employees who are able to organize their work to a certain degree by themselves as well as a management team which understands that its knowledge about the business processes is not omniscient and that allows the employees a certain degree of self-organization. Taylorism is no longer appropriate to meet the requirements of today's business world [3].

If order for process users to describe the processes they are involved in by themselves, we need a language which can be easily learned and used by traditional stakeholders, and not only by process modeling specialists.

Process models described by users in a simple language have to be executable without any programming efforts. This means that such a simple process specification language must have a formal semantics. Based on such a formal semantics, executable workflows can be generated automatically from a process model.

In spite of this fundamental simplicity, a modeling language must allow action patterns to be expressed in a compact and understandable way. This is necessary to reduce modeling efforts and to increase transparency.

Al these aspects must be supported by corresponding tools. These tools are essential for an agile company.

The following figure shows how these various aspects are mutually related, in order to get a holistic view on BPM 2.0.

Business users think in their natural language and therefore want to describe processes in the same way: in their natural language (which could explain the success of MS-Word as a tool for describing business processes).

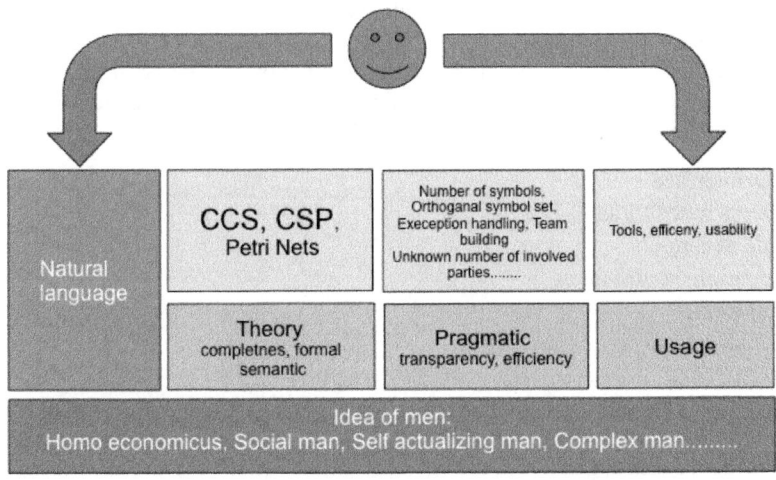

Fig. 1. S-BPM Constituents

Traditionally, in case stakeholders describe a process model, natural language is their first choice as a modeling language. Natural languages have the advantage that everybody understands them, but the disadvantage that it is impossible to derive code from them.

Formal specification methods use only a few symbols for describing models. Theoretical approaches for specification methods use the least number of symbols possible for expressing every possible process behavior. Examples for such approaches are Petri nets [4], CSP [5][6] and CCS [7][8].

Even though code can be derived from models described with these types of languages, they have the disadvantage of being abstract and therefore often difficult to understand and use by traditional stakeholders. The fact that these languages strive to keep the number of symbols for describing process behavior to a minimum makes it very cumbersome to describe complex processes. In order to reduce these efforts, some elements for description may be added. They decrease specification effort and make process models more intelligible. Although the basic description elements are sufficient to describe all processes due to Turing completeness, the addition of pragmatic modeling elements improves compactness and transparency of process specifications.

These pragmatic elements can be modified in accordance with practical experiences and target application areas. New elements can be added; others can be removed or replaced. The basic elements together with the pragmatic elements form the foundation of a specification language.

In order to support modeling methods that can be easily learned and thus widely accepted by traditional business people, these modeling languages should be closely aligned with natural languages. Such a modeling language should be based on the basic structure of natural languages, but also have a formal semantics. Formal semantics are a prerequisite for generating executable code.

This generated code includes functions that enable the production of data related to process execution, e.g., execution time and execution delays. These data represent the basis for process monitoring.

Specification languages are reflected through corresponding tools. These tools support the creation of process models and transform the models into executable code. Essentially, these tools have to support the following aspects:

- User interface
- Process repositories
- Code generators
- Execution environment
- Monitoring
- Execution analysis

4   Subject, Predicate and Object in Natural Languages meet BPM

4.1   Natural Languages

In many natural languages[1] subject, predicate and object are the basic building blocks of a sentence. The subject of a sentence is the person, place, thing, or idea that is carrying out the action denoted by the predicate. A predicate has at its center a simple predicate, which is always the verb or verbs linked to the subject [10][11].

The direct object is the person or thing that receives the action of the verb. It normally follows the verb. The indirect object is the person or thing to whom or to which the action was directed or for whom or for which the action was performed [10]. The indirect object is in a way the recipient of the direct object [11].

4.2   Business Processes

There are many definitions of the term process. The essential elements for a process are tasks, actors, resources and information [9]. These elements are combined through

---

[1] The author has checked the following languages: English, German, French, Russian, Spanish, Italian, Japanese, Chinese, Mongolian, Polish, Danish, Swedish, Norwegian.

a logical sequence of actions executed by actors using resources and/or information. This means that in process specifications the following W-questions have to be answered: "Who does what with what and when? "

4.3 Business Processes and Grammar

If you compare these W-questions with the basic structuring elements of natural languages subject, predicate and object, then you can see a close relationship.

- Subject   who
- Predicate: what
- Object: with what

The sequence of such sentences describes who executes an action on a particular object and at what time. Processes described in such a way are closely aligned with natural languages and therefore easily understood.

4.4 Objects

Subjects execute actions on objects. In object-oriented programming concepts, an object consists of data structures and operations which manipulate these data. These operations are often termed methods. This means object-oriented programming covers predicate and objects of natural languages. Although a subject executes a method on an object, subjects are not emphasized in object-oriented technologies. However, in UML actors in use case diagrams represent a kind of subject, as an exception to the general rule.

4.5 Concurrency of Subjects

Generally, several subjects are involved in a process. These subjects execute their activities in parallel. In a business process the activities of these subjects have to follow a certain sequence, e.g. a subject 'salesman' receives the request for an offer from a customer and then he or she produces an offer. For synchronizing parallel activity flows, many concepts have been developed, especially in the area of operating systems [12][13]. The exchange of messages is one of these concepts. Based on this synchronization via message exchange, Hoare and Milner developed a theory on concurrent communication systems [5][6][7][8].

4.6 Theory of Communication and Concurrency

In [8] Milner has defined a standard form for concurrent systems (p. 32 in [8], p. 68 in [7]). In a standard form, a system consists of a set of sequential agents, which run in parallel. These agents exchange messages. A business process can be seen as a standard form of a concurrent system. The sequentially executable agents correspond to the subjects. These subjects send messages to other subjects and, vice versa, subjects receive messages from other subjects.

4.7 Combining the Constituents

Based on the structure of natural languages, object-oriented programming, concurrency and communication we can build a simple language for describing business processes. This simple language has three basic types of sentences:

1. Subject_X executes operation_Y on Object_Z
2. Subject_X sends message_Y to Subject_Z
3. Subject_X receives message_Y from subject_Z

Sentence type 1 has only a simple object, whereas the sentence types 2 and 3 have a direct object (the message) and an indirect object (the receiver or sender). In the theories of Milner or Hoare the objects can be seen as simple agents and the methods as messages. This means the theories of Hoare and Milner cover this simple language in accordance with the grammar of natural languages.

Because of the underlying theory, business processes described with these types of sentences can automatically be transformed into executable programs.

4.8 Subject-Oriented BPM (S-BPM)

Subject-oriented BPM (S-BPM) is a BPM approach which focuses on the acting elements in a business process, i.e. the subjects. In many natural languages, sentences start with the subject, just as we start with the subject in S-BPM.

Subjects in subject-oriented BPM or programming are defined in conformance with their usage in grammar. Subjects are active elements in business processes and therefore the starting point of activities.

Process models and business processes can be executed at several locations in an organization, e.g. the same sales process model may be used in any subsidiary. In order to avoid describing the same process for each subsidiary in which the process is used, we use abstract subjects instead of concrete persons. For example, we use the subject "sales executive" instead of Max Huber who might be the sales executive in the subsidiary Berlin. If a process is embedded into an organization, the abstract subjects of a process are assigned to concrete persons of a target environment.

In S-BPM subjects are abstract resources which execute defined actions on objects. Subjects synchronize their activities by exchanging messages. When a process model with subjects is embedded into an organization, organizational units or persons are assigned to subjects. These subjects execute the activities as determined by the subject definition.

4.9 Other Meanings of the Term Subject

Before we continue with the development of our business process specification approach we clarify some terminology with regard to the term subject.

There are different meanings and usages of the word „subject", for example, "subject" in e-mails or "subject" in programming. "Subject-oriented programming is an object-oriented software paradigm in which the state (fields) and behavior (methods) of objects are not seen as intrinsic to the objects themselves, but rather are provided by various subjective perceptions ("subjects") of the objects." [15] Currently, subject-oriented programming is seen as a variation of aspect-oriented programming [15].

There are also "subjects" in logic where predicates are assigned to subjects. In some textbooks on the topic of logic, the term nominator is used instead of the term subject and the term predicator instead of the term predicate. [16]. However, there is no relationship to "subjects" as referred to in the grammar of natural languages.

In semantic webs "subject" has a similar meaning as in logic.

"In philosophy "subject" has a similar meaning as in the grammar of natural languages. In philosophy, a subject is a being, which has subjective experiences, subjective consciousness or a relationship with another entity (or "object"). A subject is an observer and an object is a thing observed."[17]

In conclusion, subjects in subject-oriented BPM or programming are defined conform to their usage in grammar. Subjects are active elements in business processes and therefore the starting point of activities. Subjects are abstract resources which execute defined actions on objects. Subjects synchronize their activities by exchanging messages, just as people synchronize their activities through communicating to each other.

5 BPM Modeling Approaches

In the sections above, we have shown that a process specification is based on subject, predicate and object. The various traditional modeling methods emphasize these aspects differently. Calculus of Communication Systems (CCS) and Communicating Sequential Processes (CSP) focus only on subjects. Swim lanes could also be seen as a certain type of subjects.

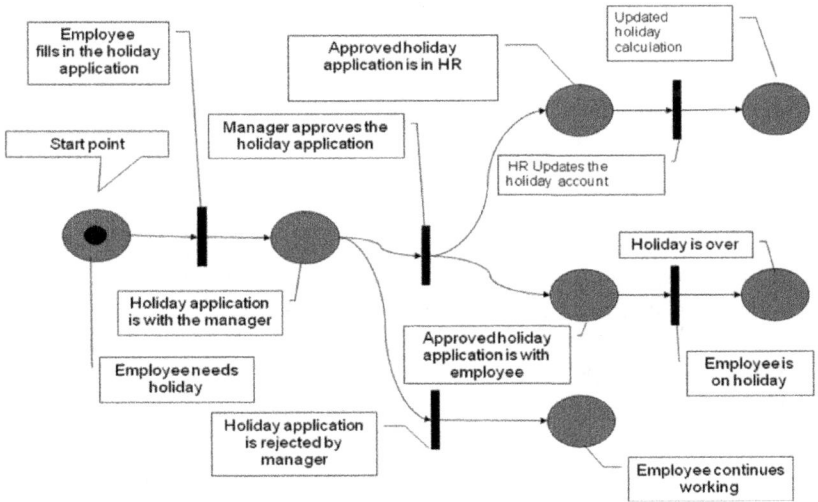

Fig. 2. Petri-net example

The dominating modeling languages of today are predicate-oriented. Petri nets as a theory for concurrent systems focus on the activities. The transitions in Petri nets correspond to the predicates, which are executed. Figure 2 shows a Petri net describing the application for a holiday.

In Petri nets subjects and objects must be added as comments, since they are not part of the model itself.

Variants of flow charts are often used for specifying processes. An example of such a flow chart based method is event driven process chains (EPC) [14]. EPCs are part of the ARIS (Architecture of Integrated Information Systems) architecture (ARIS-House). The four aspects of business process specifications considered in ARIS can be described as follows: We find the aspect "When" at the center of the ARIS modeling framework which therefore represents the control flow. The aspect "predicate" ("Which?") is found in the functions, the aspect "object" ("What?") in the data and the aspect "subject" ("Who?") in the organization. Figure 3 depicts the ARIS architecture and shows which constituents cover the various aspects of business process specifications.

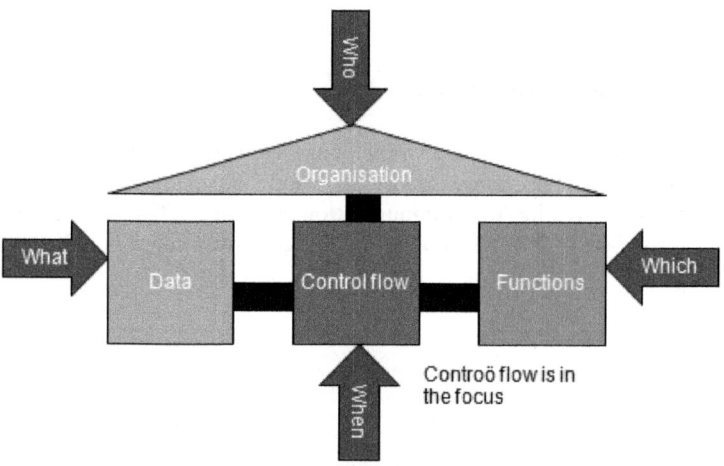

Fig. 3. Building blocks of ARIS in context of 4 W's

The third possible starting point for business process specifications can be the objects. With such an approach you start with an object, such as an order, an invoice, etc. and the modeler specifies the changes on that object.

An example of such an approach is described in [18] termed Artifact-Based-Transformation. In this process modeling technique, key business artifacts are identified and their life cycle is traced when they are processed by the business.

6 Subjects Define the Granularity of the Actions in Business Processes

We have shown that a business process specification consists of three major aspects. When defining a process specification, modeling can start with the actions, the subjects or the objects. The starting point depends on the focus of the applied specification approach.

The "traditional" starting point for modeling a business process is actions. However, if you begin modeling with actions, determining the appropriate granularity for the actions defined in the process can be difficult. For example, you may want to model a simple order process as shown in the figure below. Which level of detail should be chosen?

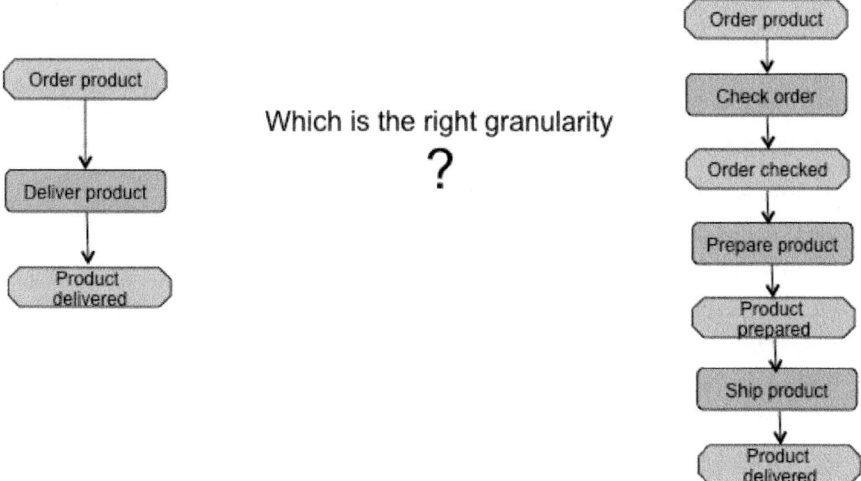

Fig. 4. Specifications with different degrees of granularity

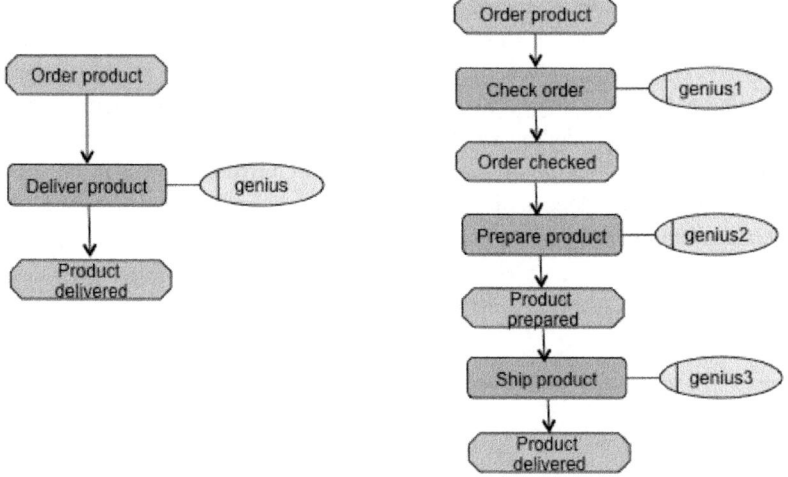

Fig. 5. Assignment of actors (organizational units)

The only way to decide which is the most appropriate is by introducing the subject to the model. If there is only one subject, let's call him genius, who is assigned the action "Deliver product", then the granularity on the left fits perfectly. On the other hand, if there is more than one subject involved, let's say genius 1 to genius 3, the modeler has to split the action into sub-actions whereby each sub-action is processed by no more than one subject.

Therefore, the subjects define the granularity of the actions of a business process which means that modelers should always start modeling a process by specifying the subjects, and not the actions. This is independent of the technology being used.

If process modeling starts with activities or objects, the modeler may have to adapt the process granularity to the subjects executing the activities.

The S-BPM approach for modeling business processes consists of the following steps:

1. Identify business processes and create a specific process network
2. Identify the subjects in a process
3. Identify the messages exchanged between subjects
4. Identify the payload of the messages (information objects exchanged)
5. Define the behavior of each subject (send, receive, Execute)
6. Embed the model in the respective environment (Organization and IT)

7 S-BPM Modeling Language

Based on the considerations above, we have developed a subject-oriented approach for specifying business processes. This approach is called subject-oriented because modeling is centered on the acting elements in a process, i.e. the subjects.

The following table gives an overview of this S-BPM language called PASS (Parallel Activity Specification Schema). It is based on the grammar of natural languages, Hoare´s and Milner´s theory and the concepts of object-oriented programming. These basic concepts are extended by additional pragmatic elements, which allow a compact and transparent specification of various recurring behavioral aspects in business processes. We have also added additional elements to support the structuring of highly complex and extremely large process systems.

Symbol	Name	Description

This table gives a short overview about the concepts of subject-oriented modeling. It does not show all the details which may be required to model certain business cases.

Basic Modeling

The basic modeling features theoretically cover all the requirements for modeling business processes

Communication Structure

In the communication structure it is defined which subjects participate in a process and which messages they exchange.

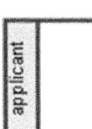

Subject

The acting elements in a process are called subjects. Subjects are abstract resources which represent acting parties in a process. Subjects are process specific roles. They have a name, which gives an indication of their role in the process. Subjects send or receive messages to or from other process participants (subjects) and execute actions on business objects. The sequence in which they carry out these activities is described in the subject behavior (see below).

Message

Subjects exchange information and synchronize their activities by exchanging messages. Each message has a name. The message name should give an indication of the content and purpose of a message.

Messages can be exchanged asynchronously or synchronously. Each subject has an input pool in which the sending subject deposits the messages for this subject. The corresponding subject accepts messages by removing them from the input pool. By configuring the input pool of a subject it can be specified which messages from which subject are received synchronously or asynchronously.

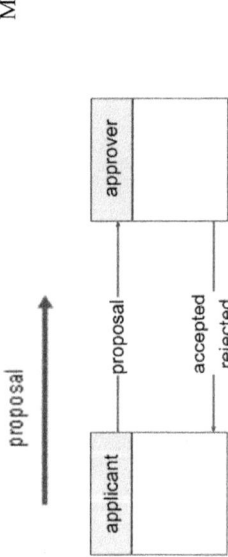

Business Objects

Messages transport business objects. Each business object has a name and content. The name of a business object should give an indication about the purpose and content of this business object. Business objects can be orders, invoices a.t.l.. The business objects contain information. This information can be hierarchically structured. Business objects are exchanged between subjects via messages. Business objects are the payloads of messages. Messages with different names can transport different business objects.

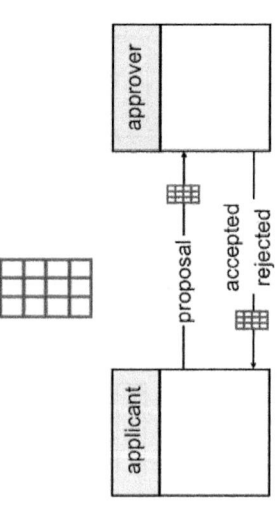

Behavior

The behavior of subjects describes in which sequence messages are sent or received and which actions are executed at what time.

Send messages

Subjects send messages to other subjects. In a send state, a subject tries to send several messages alternatively. In our example, the subject tries to send either message M1 to subject S1 or message M2 to subject S2. If a message can be sent to the addressee, the corresponding transition is executed. If both messages can be sent randomly, one message is chosen and sent. Messages cannot be sent if the addressed subject only accepts the message synchronously and is not in a state in which this message is expected.

Receive Messages

Subjects receive messages from various other subjects. In a receive state a subject can accept several messages alternatively. In our example, a subject waits for the message "rejected" or "accepted" from the subject "approver". If one of these messages is available, the corresponding transition is executed. A message is available if it is already in the input pool or it is directly offered by the sender. If both messages are available, one of them will be selected randomly.

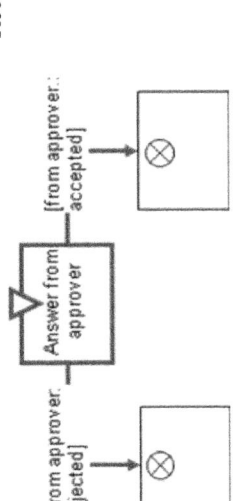

| Actions /Activities | Subjects also execute internal actions. Internal actions are defined on internal data like business objects. Internal actions can change business objects or check the values of elements in a business object, or both. The action is executed in the state. The execution of an action can end with several results. In our example, the action "Check Proposal" can have the result "accept" or "reject". |

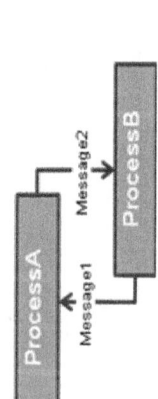

Advanced Modeling
Communication Structure

In order to define complex process systems, different modeling methods are added. These features do not extend the power of subject-oriented modeling (this is not possible as a result of Turing completeness) but help to describe complex processes in a more transparent way.

| Connected Processes | If processes become too complex, e.g. too many subjects, then processes can be split into several connected processes. Processes are connected via the communication of subjects in the connected processes. Subjects in one process exchange messages with subjects in other processes they are connected to. From a process perspective, subjects in connected processes which interact with each other are called external subjects. |

Each process can be connected to several other processes.

This allows the creation of networks of processes.

The example shows two connected processes: ProcessA and ProcessB. Subject2 in ProcessA interacts with Subject3 in ProcessB. Subject2 in ProcessA and Subject3 in ProcessB exchange the messages "Message1" and "Message2". Subject3 in ProcessB is represented as an external subject in ProcessA and subject2 in ProcessA is represented as an external subject in ProcessB (cross reference).

Service processes are very similar to connected processes. In a service process there is a subject which is visible to all other processes. This subject is called the interface subject. This interface subject accepts messages from subjects in any other process. These subjects are unknown to the interface subject of a service process.

Messages to an interface subject are service invocations. An interface subject accepts messages and stores the name of the sending subject in a variable. The variable is used to send messages back to the invoking subject.

In service processes there is no cross reference between connected processes. A service process does not know which subjects in which processes send messages to it.

The example shows how a service process is used. Subject3 of the service process is visible to all invoking processes.

Service Processes

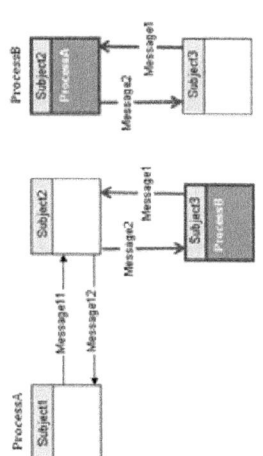

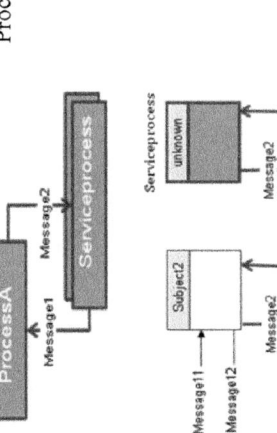

Subject2 in ProcessA sends "Message2" to Subject3 in the service process "Serviceprocess". Subject3 accepts the message "Message2" from any subject in any process. In order to send messages back to the invoking subject, Subject3 in the service process stores the sender information in a so-called address variable.

Multi-processes

Multi-processes are an extension of connected processes. If a multi-process receives a certain message from a connected process, a copy of the multi-process is generated. Every time this message is sent to the multi-process, a new copy is generated. In our example, every time subject2 in ProcessA sends the message "Message2" to the subject3 which is part of multi-process "MultiprocessB", a new copy of "MultiprocessB" is produced.

Multi-processes allow superficial modeling of certain business cases. Examples of such cases might be reviews of papers or request for proposals with an unknown number of reviewers or suppliers.

In the example, a copy of "MultiprocessB" is created when the message "Message2" is sent to Subject3 in "MultiprocessB". In Subject2 in ProcessA the messages "Message1" from the various copies of "MultiprocessB" are received. A business object can be transported in each copy of "Message1". These business objects are stored in a

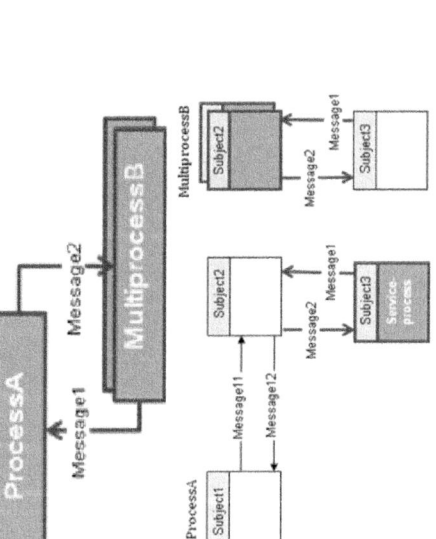

business object array contained in Subject2 in ProcessA.

Hierarchical
Process Network

A process can be connected to another process. This connected process can in turn be connected with other processes and so on and so forth. In this way, hierarchical networks of connected processes can be constructed.

Our example shows a network with three connected levels of processes (ProcessA, ProcessB and ProcessC). Each process can consist of subjects and other processes. ProcessA consists of the processes ProcessA1 and ProcessA2. ProcessA1 and ProcessA2 are also connected. In our example, we show only the subjects in ProcessA, which are visible to subjects in other processes (external subjects).

There can be any number of internal subjects, which are not visible to other processes.

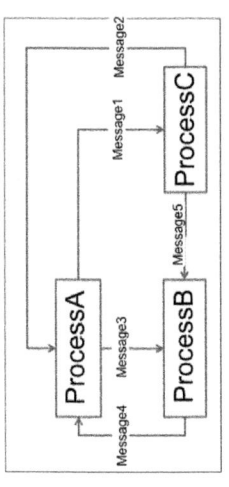

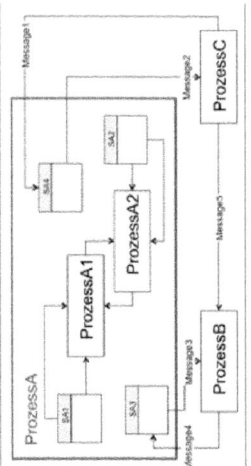

Behavior

The following constructs allow the description of certain behavioral issues in a transparent way.

If it is necessary to accept a certain set of messages in many states, e.g. an order, which can be cancelled in any delivery state, then a behavior specification can become very complex. Instead of adding the corresponding transitions to any affected state, the handling of these messages is separated from the main activity sequence (or the so-called "happy path" which is referred to in this example as the standard or default scenario, not considering any exception or error handling). Messages, which are not handled in the happy path, are called exception messages. These messages are handled in separate behavior specifications, or so-called message observers. A subject can have several message observers. A message observer can handle a set of observed messages and it can be valid for several states in the happy path. If a subject is in a certain state and an exception message arrives and this message is observed in this particular state, the affected subject jumps into the corresponding observer behavior. After execution of the observer behavior the subject continues on in the happy path. In which state the subject continues is defined in the observer behavior.

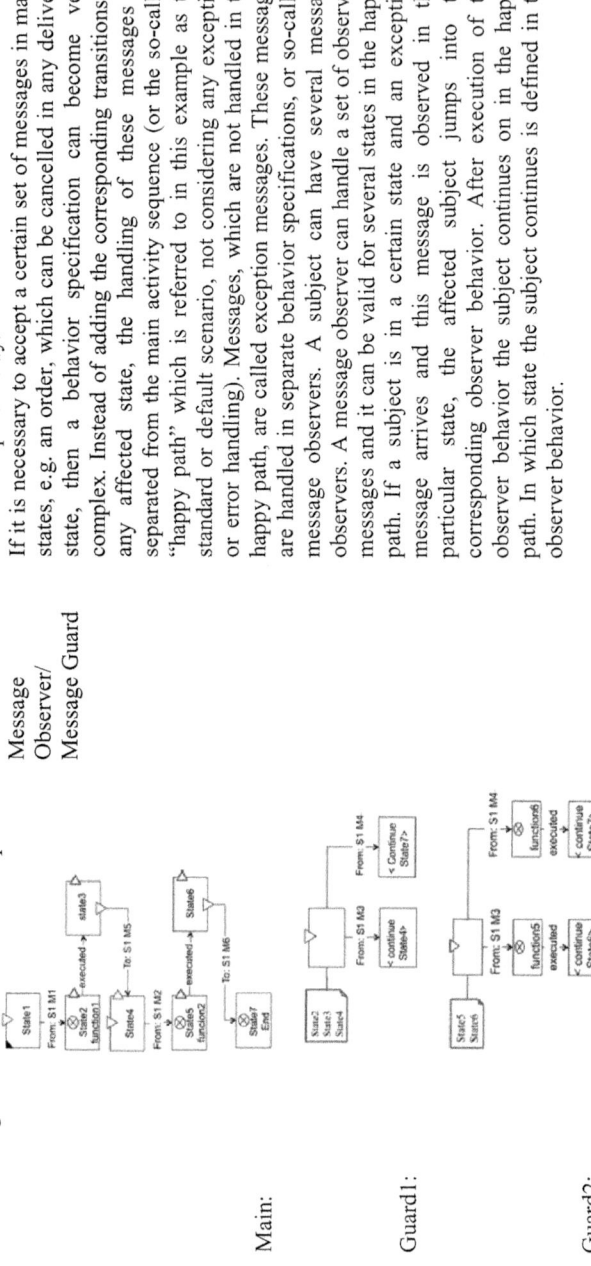

Message
Observer/
Message Guard

Main:

Guard1:

Guard2:

In our example, two different observers (guards) handle the messages "M3" and "M4" from subject S1. In one observer (guard1) the states "state2", "state3" and "state4" are observed and in the second observer (guard2) the states "state5" and "state6".

If one of the observed messages is available in one of the observed states, the subject continues with the corresponding observer. At the end of each observer, it is defined in which state of the main behavior the subject continues its execution.

Choice/
Multipath

If it is not necessary that specific actions are executed in a certain order, a choice operator can be used to express the overlapping execution of action sequences. Normally, each subject has only one active state. An active state is the state representing the next action executed by the corresponding subject.

Our example shows such a choice operation. The open choice operator defines the number of overlapping sequences. In our case, there are three overlapping sequences. Each of these sequences has its own active state. After the "Start choice" operator the next states are the initial active states of the choice clause. A subject can choose randomly which of the active states will be executed next. This means during the execution of a choice clause the subject can jump between the various active states.

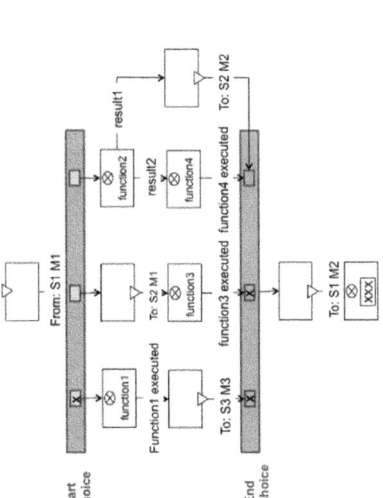

Some sequences in a choice operator must be executed (in our example the left path, marked with an x at the start and end of the sequence). Others may be executed, but if execution has started they must be finished (in our example the sequence in the middle, marked with an x at the end of the sequence). Still others may be executed, but not necessarily finished (right sequence, no marks).

In some cases, identical activity sequences are used in behavior specifications of a subject at several places in the behavior description or in several different subjects. These sequences can be defined once as a so-called macro. Macros can be embedded in subject behaviors where the corresponding macro behavior is required.

In our example, we have defined a macro in which a message is sent to a subject and two answers are expected. After the receipt of each answer, a corresponding follow-up activity is executed. This macro can be embedded in the behavior of any subject, provided the partner subject is the same (in our case it must be subject S1). Hence, macros can be only used in a very restrictive way.

In order to define macros independent from communication partners it is possible to define macro types. In macros, so-called formal subject names are used for specifying senders or receivers of a message. Embedding of a macro type means that the formal names are replaced by actual subject names depending on the embedding in a subject behavior.

Macros

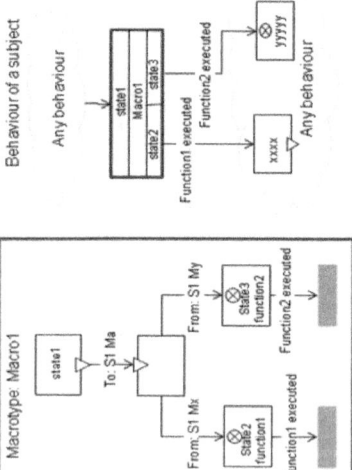

8 Organizational and Technical Implementation

The modeling elements below allow the definition of business processes independent of their organizational and IT environment. The models are abstract. The subjects represent the acting elements and the business objects the required information. In a succeeding step, process models are embedded in an org chart and IT Infrastructure. Embedding a model in an org chart means, you assign subjects to organizational elements. Embedding an org chart in an IT Environment modelers have to implement the following:

- the operations on business objects and how these objects are stored in files and/or data bases,
- the integration of existing applications,
- the user interfaces for human interactions,
- the way messages are exchanged,
- the way behavior of subjects is controlled,
- and the storage of the monitoring

9 Summary and Conclusions

In this article we started with properties of BPM 2.0. BPM 2.0 supports agility, simplicity and direct execution of business process models. In order to meet these requirements, several concepts are combined. The basic elements subject, predicate and object of the grammar of natural languages represent the starting point. The reason we use the structure of natural languages as a starting point for defining a language for describing business process models is the fact that people are generally familiar with the natural language (or languages) they use to communicate with others. Based on this structure, concepts in computer science are identified which allow us to build a formal language with a structure in accordance with natural languages. CSP and CCS from Hoare and Milner are used to represent subjects, and the concept of object-oriented programming to represent predicate and object.

Based on this basic structure a modeling language is derived which allows the specification of complex process systems.

The embedding of process models is only outlined. Detailed publications related to the implementation aspects are in preparation.

Many aspects of the described modeling concept are supported by a tool suite developed by jCOM1 [19]. This tool set allows the specification and execution of complex business processes.

Further development is necessary for covering control and monitoring aspects.

References

1. Silver, B.: BPMS Watch: Make Way for BPM 2.0, Bruce Silver Associates (Monday March 27, 2006),
   http://www.bpminstitute.org/articles/article/article/
   bpms-watch-make-way-for-bpm-2-0/news-browse/1.html
   (last access March 2010)

2. Roychowdhury, P., Dasgupta, D.: Take Advantage of Web 2.0 for Next-generation BPM 2.0, http://www.ibm.com/developerworks/webservices/library/ wsweb2bpm2/index.html#author2 (last access March 2010)

3. Kurz, M.: BPM 2.0, Organisation, Selbstorganisation und Kollaboration im Geschäftsprozessmanagement. forFLEX-Bericht Nr.: forFLEX-2009-002 Bamberg, Erlangen-Nürnberg, Regensburg (2009)

4. Reisig, W.: Petrinetze, Eine Einführung. Springer, Heidelberg (1990)

5. Hoare, A.: Communicating Sequential Processes. Communications of the ACM 21(8), 666–677 (1978)

6. Hoare, A.: Communicating Sequential Processes. Prentice-Hall, Englewood Cliffs (1985)

7. Milner, R.: Communication and Concurrence. Prentice-Hall, Englewood Cliffs (1989)

8. Milner, R.: Communicating and Mobile Systems: The Pi-Calculus. Cambridge University Press, Cambridge (1999)

9. Fischermanns, G.: Praxishandbuch Prozessmanagement, Götz Schmidt, Gießen (2006)

10. Spark Charts: ESL Grammar, A Barnes and Nobles Publication (2004)

11. University of Ottawa, Subject and Predicate, http://www.writingcentre.uottawa.ca/hypergrammar/ subjpred.html (last access March 2010)

12. Hansen, P.B. (ed.): The Origin of Concurrent Programming. Springer, New York (2002)

13. Fleischmann, A.: Distributed Systems – Software Design and Implementation. Springer, Heidelberg (1994)

14. Scheer, A.W.: ARIS – Vom Geschäftsprozess zum Anwendungssystem. Springer, Heidelberg (2002)

15. Wikipedia, http://en.wikipedia.org/wiki/Subject-oriented_programming (2010)

16. Detel, W.: Grundkurs Philosophie, Band 1 Logik, Reclam, Stuttgart (2007)

17. Wikipedia, http://en.wikipedia.org/wiki/Subject_(philosophy) (2010)

18. Chao, T., et al.: Artifact Based Transformation of IBM Global Financing. In: 7th International Conference on in Business Process Management, BPM 2009 (2009)

19. jCOM1, http://www.jcom1.com

Case Study: The Process Portal – Process-as-a-Service Central Platform for Work-, Information- and Knowledge Processes in the Company

Anton Kramm

Valial Solution GmbH
Hettenshausener Straße 3, 85304 Ilmsmünster, Germany
A.Kramm@valial-solution.com

Abstract. This case study shows how S-BPM in combination with other technologies is helpful to increase the agility in software development and implementation. As processes become more agile, companies become more agile. From the perspective of IT-services orchestration and adaptability play a crucial role. From the perspective of the organization of work rules, patterns of behavior and events for dynamic process execution are key enablers of velocity.

Keywords: agility, process portal, service orchestration, dynamic execution, knowledge work.

1 Motivation

Why do we think S-BPM will play a crucial role in the change of software industry?

The first reason concerns the rate of market transparency. It has changed in a way that all parties (need to) know information relevant for their own business, in combination with the permanently increasing globalization demands for higher flexibility and agility in enterprises. This means the time to react and to adjust to new market conditions has to be reduced considerably. However, in most of the cases IT is not flexible enough and its representatives do not understand the needs of the business experts.

Secondly, a company is only growing if it is the number one in a certain market. Today, growth is mostly only possible by superseding. In particular, companies have to be faster than their competitors.

Thirdly, employees become more and more knowledge workers. As such they represent value assets of the enterprise. Companies must support them, not only in terms of IT-tools, but also in the proper way with information: in high quality just in time (i.e. at the proper step of the business process).

Being part of a changing world enterprises face changing business requirements. Today, in most cases IT is too inflexible to meet continuously changing requirements. It often takes months to change an IT-system and when it has finally been adapted it is not up-to-date. Therefore, there is a need for efficient, customizable IT assistance.

Taking a look at big companies they have all (nearly) the same aims: more production and sales, increasing the return-on-investment, employee retention and - most

H. Buchwald et al. (Eds.): S-BPM ONE, CCIS 85, pp. 107–114, 2010.

important - customer satisfaction. All products and services have to create value for the customer, and there is no gap between this business side and IT. The gap only exists in the company itself. Consequently, companies need to grow without expanding which translates to optimizing their processes.

2 Business Requirements to Process Portals

What are the requirements to a process portal from a business side in the course of optimizing business processes? First of all, it is the provision of a "cockpit" for all employees where a certain work processes can be modeled, administrated, planned and controlled comfortably by the employees themselves. It should have just one front-end for all users and reflect the status of all processes available in real-time. Then, management does not need to train and support employees on all the different tools and front-ends involved in the business processes. S-BPM and portal technology enable single front-end to all the applications.

Moreover, all analysis tools, decision making- and planning instruments need to be interlinked according to respective responsibilities and support both - managers and employees - in their tasks and working activities.

A process portal should be the central portal platform for corporate-, business- and work-processes to make them faster, leaner-centered and more efficient: Information for the right person, in the right place, at the right time, and in the right form. This should be the central contribution of IT to the company's success, enabled by novel technologies and knowledge media.

A portal is an intermediate between companies, employees and people between processes, technology and knowledge (see Figure 1). In conventional BPM-platforms or SOA-infrastructures knowledge about services or information is missing. S-BPM could help to bridge the gap between the information being available and the compliance.

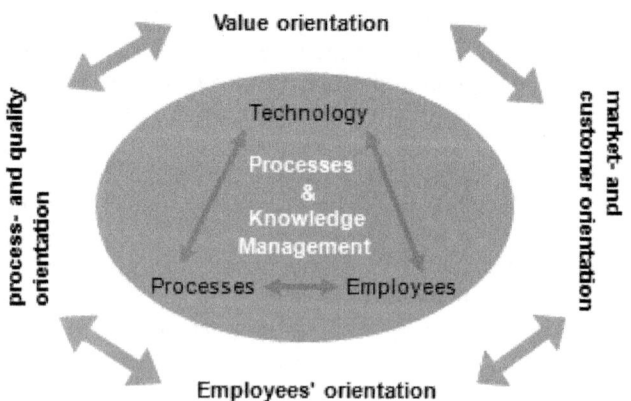

Fig. 1. Processes, knowledge and technology triangle

This is how Valial Solution operates. Our business model does not include management or traditional process consulting in which a somehow "best" business process is designed that has to be implemented by somebody else. Valial Solution offers process implementation in an innovative way (see Figure 2). The solution allows the customer to use "processes as a service" based on the company's infrastructure and standards. We provide consultancy results as process containers. A process container is reusable. It is customizable with S-BPM, and can be tailored to the context of a company while the core process remains serves as focal point. A customized process can be orchestrated, which means that it is integrated in the company's existing system infrastructure (services and tools - see Figure.2).

The benefit of using process containers is (i) they are secure because they are tested in other real environments, and (ii) they are already in operation. There is no go-live-risk from a newly designed application to an operating system like in traditional software projects.

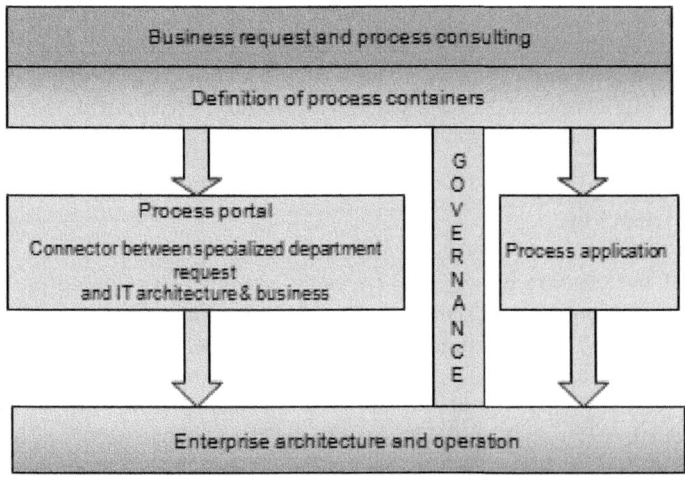

Fig. 2. Process container in a company

3 Implementation Issues

The challenge for IT is to bridge the gap between business requirements and the underlying IT system infrastructure in an ever changing environment, and therefore ever changing requirements. This is called the "conversation gap". In conventional software and BPM systems this gap can only be bridged if (at all) the company's software stems from a single vendor or IT provider. One will hardly find any company depending on a single vendor or IT provider.

With S-BPM enterprises have the possibility to bridge this gap – see Figure 3. The model includes the process flow defined by the line of business, and also the communication between all process parties. The portal provides the required connectivity and

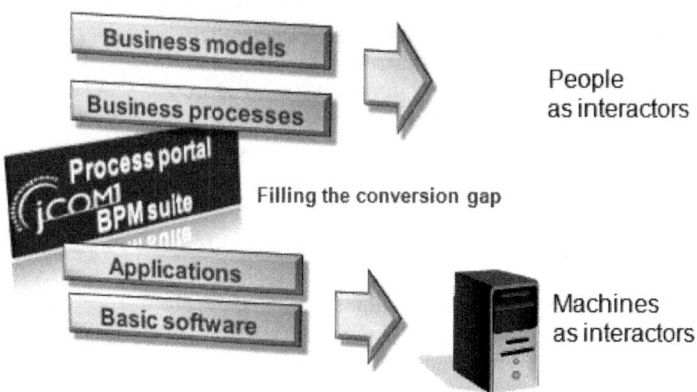

Fig. 3. Bridging the conversation gap with S-BPM

interaction between backend systems. Utilizing S-BPM, organizations have the possibility to model and orchestrate in a single step by letting the business users specify their needs. The result is a workflow and a new composite application, with the business logic in the model, rather than in the code.

4 SOB Lifecycle

In Figure 4 the solution process and lifecycle is shown. It is based on open-source standards but not on notation standards. The first step is to generate the process model (BPM Modeling). Then, this process is orchestrated (Service-Choice) with respect to the existing infrastructure. In order to maintain and evolve these services a meta-process is required which should be based on a standard like CMMI (Capability Maturity Model Integration) or SPICE (Software Process Improvement and Capability Determination). This helps the user in the line of business to get the services they need without knowing who in IT is responsible. Users simply start the (meta-) process to receive their services. After the orchestration the result has to be validated (Validation). This is not only business validation, but also operation validation. It is checked whether the process can run on the existing infrastructure (e.g., whether the services are able to handle the required amount of requests). Therefore, the ITIL (Information Technology Infrastructure Library) check is integrated in the validation step. After the validation the users can upload the process and operate it. Finally, the running process is evaluated. Note that it is not evaluated before operation since it could be speculative without data from actual process instances.

Business Activity Monitoring (BAM) based on Complex Event Processing (CEP) monitors the life process in real-time. The generated data of these events allow optimizing the processes. Consequently, there are more opportunities to start a process, as it can be started by any event, e.g., by fraud or fault, and not only by human interaction. Users might run a simulation in case they are not sure which process is the correct one to initiate (see Figure 4).

Fig. 4. S-BPM cycle and validation

5 Examples

In the following some examples of implemented process portal solutions based on S-BPM are shown.

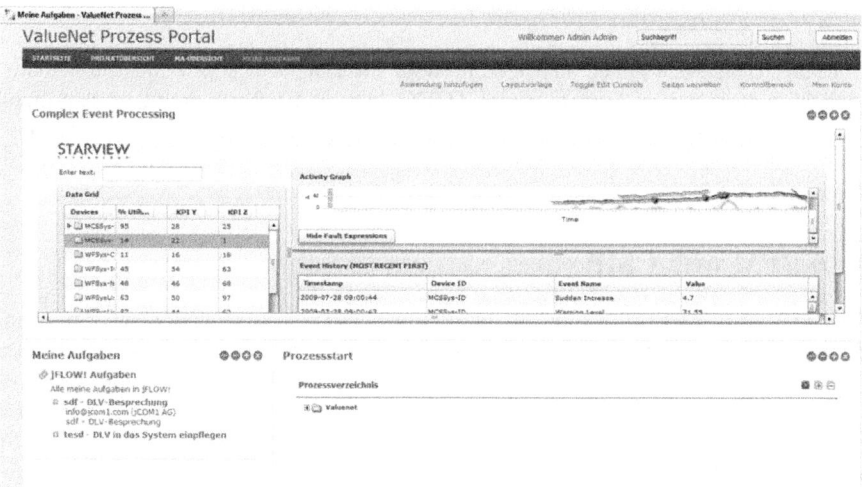

Fig. 5. ValueNet HR portal

Figure 5 show a screenshot of the ValueNet HR portal. Based on S-BPM process models, customers and the company are able to handle HR processes addressing tasks, data in an intranet in a highly flexible way. The processes are monitored by the CEP engine-provider Starview and comprise a personalized management cockpit. This solution gives the customer also the possibility to react directly to faults or events like an expert system.

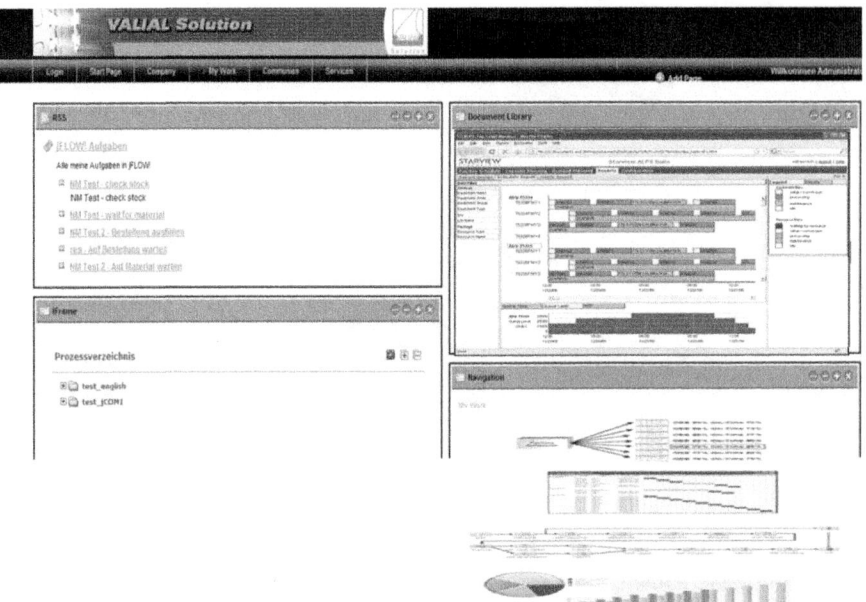

Fig. 6. VALIAL Intranet with BI dashboards and simulation

Figure 6 shows another VALIAL intranet solution, which provides immediate information in terms of "my personal TODO list" and reports about project statuses providing controlling checks against targets. The simulation portlet shows the resource planning based on orders, human resources, available servers and possible changes (what-happens-if scenarios due to events occurred)..

6 S-BPM with CMP

The goal is to integrate a more dynamic business process management. This can be done using the events from the process instances to identify event patterns from this actual data and deduced key performance indicators (KPIs).

Figure 7 gives an overview on how the different technologies are integrated and how the required exchange of data exchange can be realized. The process portal, Starview CEP, and jCOM1!'s S-BPM Suite directly transfer and monitor data in real time. In this way the users are offered decision portlets based on events or processes, control KPI, start escalation, or based on automatic decisions, the initiation of a new process instance.

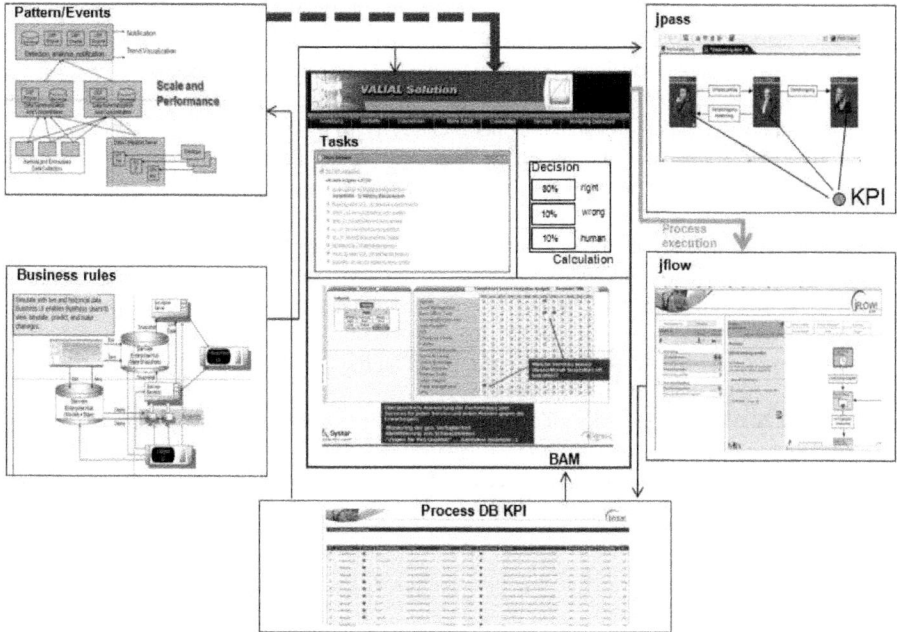

Fig. 7. Overview diagram CEP - Portal - S-BPM

7 Conclusions

The goal is: IT as a business enabling function. The requirements to be met are:

- High flexibility and automation in the provisioning of IT services
- Customer adds services to his process models and orchestrates them to compound applications
- Costumer finds the services and processes in the repository
- Customer defines rules, patterns and events for dynamic process execution
- Customer simulates different event scenarios and decisions
- IT provides services and monitors the availability (ability to deliver)
- IT checks the reusability and sizing of services with ITIL processes
- IT places end-to-end monitoring to review the SLA-retention
- Documentation and logging of the real processes (retention of compliance)

Therefore, the business process defines IT-solution!

References

1. Chakraborthy, D.: Extending the reach of business processes. IEEE Computer 37(4), 78–80 (2004)
2. Fleischmann, A.: What is S-BPM? In: Buchwald, H., et al. (eds.) S-BPM ONE. CCIS, vol. 85, pp. 85–106. Springer, Heidelberg (2010)

3. Heackel, S.S.: Adaptive enterprise: Creating and leading sense-AND-respond organizations. Harvard Business School Press, Cambridge (1999)
4. Havey, M.: Essential Business Process Modeling. O'Reilly, Beijing (2005)
5. Laudon, K.-C., Laudon, J.P.: Essentials of management information systems: Managing the digital firm, 6th edn. Pearson, Upper Saddle River (2005)
6. Lewis, M., Young, B., Mathiassen, L., Rai, A., Welke, R.: Business process innovation based on stakeholder perceptions. Information Knowledge Systems Management 6, 7–17 (2007)
7. Rouse, W.B. (ed.): Enterprise transformation: Understanding and enabling fundamental change. Wiley, Hoboken (2006)
8. S-Cube Consortium: Survey on Business Process Management (2008), http://www.s-cube-network.eu
9. Spurway, K.: The State of BPM: Perspective of an Industry Insider, http://www.bpm.com (10.2.2010)
10. Scheer, A.-W.: ARIS - Modellierungsmethoden, Metamodelle, Anwendungen, 4th edn. Springer, Berlin (2001)
11. Stephenson, S.V., Sage, A.: Architecting for enterprise resource planning. Information Knowledge Systems Management 6, 81–121 (2007)

Case Study: AST Order Control Processing

Gabriele Konjack

Vice President Order Processing & Accounting
Finanz Informatik Technologie Service
Einsteinwg 17, 85609 Aschheim bei München
www.f-i-ts.de

Abstract. Even process-driven organizations need a roadmap when switching from traditional to actor-driven BPM. The case described in the contribution reveals significant organizational changes, both, when organizing a BPM project, and when implementing communication-based processes. Once the lean approach has become part of the mental model, economic benefits can be expected due to the increase in process quality.

Keywords: order control, organizational change, subject orientation, case management.

1 Introduction

In the following a business process management project from a finance service provider (FITS) is described. FITS (Finanz Informatik Technologie Service) offers IT services for German banks, especially saving banks. The goal of this project was to define and implement the order control process of the FITS. After introducing the control process we report on the project setting, the actual procedure, and the impact on the organization. Finally, we sketch some lessons learnt in the course of this initial S-BPM project of FITS.

2 The Order-Control Process

The order control process describes the processing of service orders from the various customers. Customers order new IT services, such as an additional Web Server or database server. If the order is accepted by FITS an instance of the order control process is initiated. The order is handled according to the corresponding process instance.

The following figure shows an example of the execution of the order control process, namely how an order for installing and configuring a server system is processed.

A client or customer sends the order to the FITS. The order is accepted by the international sales and distribution organization. Then the order is forwarded to the order control and processing unit. In that unit TAVs ("Technische Auftragsverantwortliche" or Technical Order Managers) coordinate the realization of an order. They divide an order into work packages, establish teams and direct the corresponding

H. Buchwald et al. (Eds.): S-BPM ONE, CCIS 85, pp. 115–120, 2010.
© Springer-Verlag Berlin Heidelberg 2010

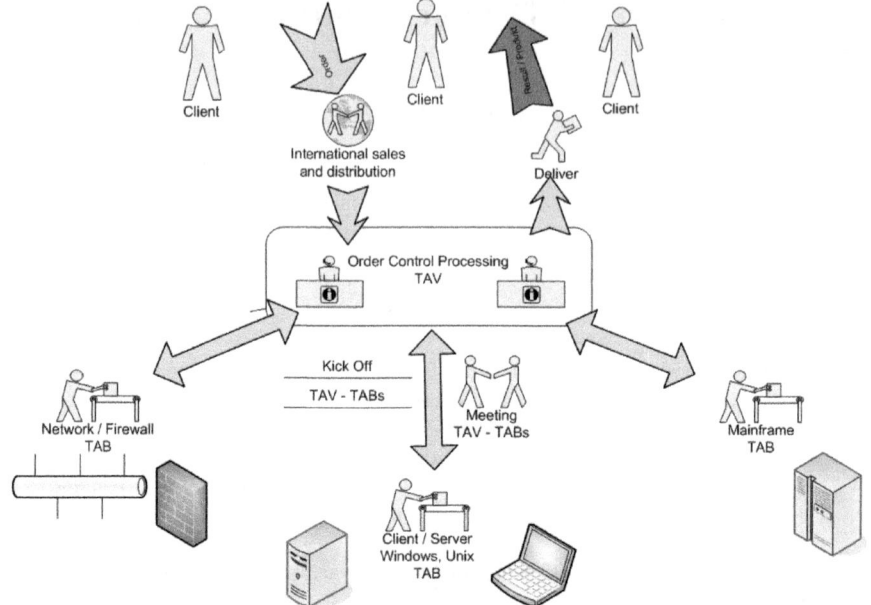

Fig. 1. Processing a server-system order

work packages to the right teams. The teams are managed by so called TABs ("Technische Auftragsbearbeiter" or Technical Order Processors). The TABS are responsible for processing the work packages on time, within budget and with the required quality.

3   The Order-Control Process Project

A project was established with the general goal of improving the order control process as a whole. In order to do so the following various aspects of the process with improvement potential were identified:

- Optimization of order processing
- Monitoring in conjunction with adherence to delivery dates, time and efforts (= costs) through central coordination and governing
- Reporting (ongoing) with respect to the status of each order being processed

In order to improve the transparency of the execution of process instances the following aspects need to be improved:

- Quality assurance of order processing regarding content, time and costs
  - Kick off Meeting, order tracking, remaining functions, reporting of order status.
- Communication interface between sales/customer and processing units
  - Detection of troubles and problems and initiation of escalations

- Detection and approval of the time allocations in the SAP System CATS
- Plausibility check of order/contract and technical implementation
- Reporting overviews covering all orders with regards to numbers, content, time and implementation status

As several organizations are involved in the order-control process, the project team included representatives from each of these organizations. The following centers of competence participated in the project:

- Network and firewall
- Windows-based server and clients
- Unix-based server incl. SAP
- Mainframe and (banking) applications

One key user responsible for tools and methods as an internal service provider was also a member of the project team. His areas of responsibility within the project included:

- Optimization of methods and tools
- Quality management

In the course of the project it turned out that this investment of person power and time is very useful and profitable.

4   Why Use the jCOM1 Approach?

The initial situation at FITS has been as follows:

- SAP CS (only used in division „order processing"):
  - Used to collect all data (order, time limits, work package, order status, contact persons, e.g., TABs …)
  - Huge amount of manual data collection
  - Technical interface to SAP SD
- MS Office mails and/or MS Outlook tasks
  - Communication between TAV and all other contact persons
- Existing and available tools within the company for „workflows" required a major amount of programming effort (time and costs)

Consequently, the management was looking for a tool which would

- be new for all members of the process,
- allows to replace mails, outlook tasks and SAP CS, in particular
  - enables each member of the process to document his/her own responsibilities and status, and make this information visible to everyone else
  - provides an interface to SAP (SD)

The tool options considered were

- ARS and SAP Workflow
- A new tool

FITS has had extensive experiences with the first of these options and they have proven to have significant drawbacks. Therefore, the management decided to use a new tool and chose the S-BPM Suite from jCOM1. It seemed to be easy to use and supported the straightforward integration of all involved parties. In particular, it seemed to allow meeting major requirements, namely

- replacement of mails, outlook tasks and SAP CS
- interface to SAP (SD)

Finally, the new workflow tool should make the execution of the order control process more transparent for management.

5   Order Processing Means Communication

The processing of an order is nothing else but communication between all involved parties.

- Governing a (small) project, a contract, an order is communication
- To communicate you have to talk to one another in a special sequence
- Order →Initialization → Work in progress → Ready for dispatch → Acceptance
- You provide information
  - You ask for status
  - You get status information
  - You give status information and comments to other persons

S-BPM is communication-centered and, thus a method which was well-suited to meet the requirements.

6   Implementation Experiences

Using jFLOW!, the workflow component of jCOM1's S-BPM Suite, the company was able to initially implement the first complete process in only a few days. In this way a rapid decision was made to use jFLOW! as the tool of choice for the new „order processing" workflow.

When testing started the developers recognized that the process they thought they knew so well was much more complex than they realized, and that they needed to add some additional features, among them:

- Database operations for information transport and (internal) auditing
- Fluctuating members of the process depending on the specific order
- Short cuts for easier usage

Consequently, the final implementation took much longer than initially expected.

7   Supporting the Order-Control Process

The following figure shows the various subjects in the order control process of FITS and the communication between them. The boxes AM, TAV, TL and TAB are the

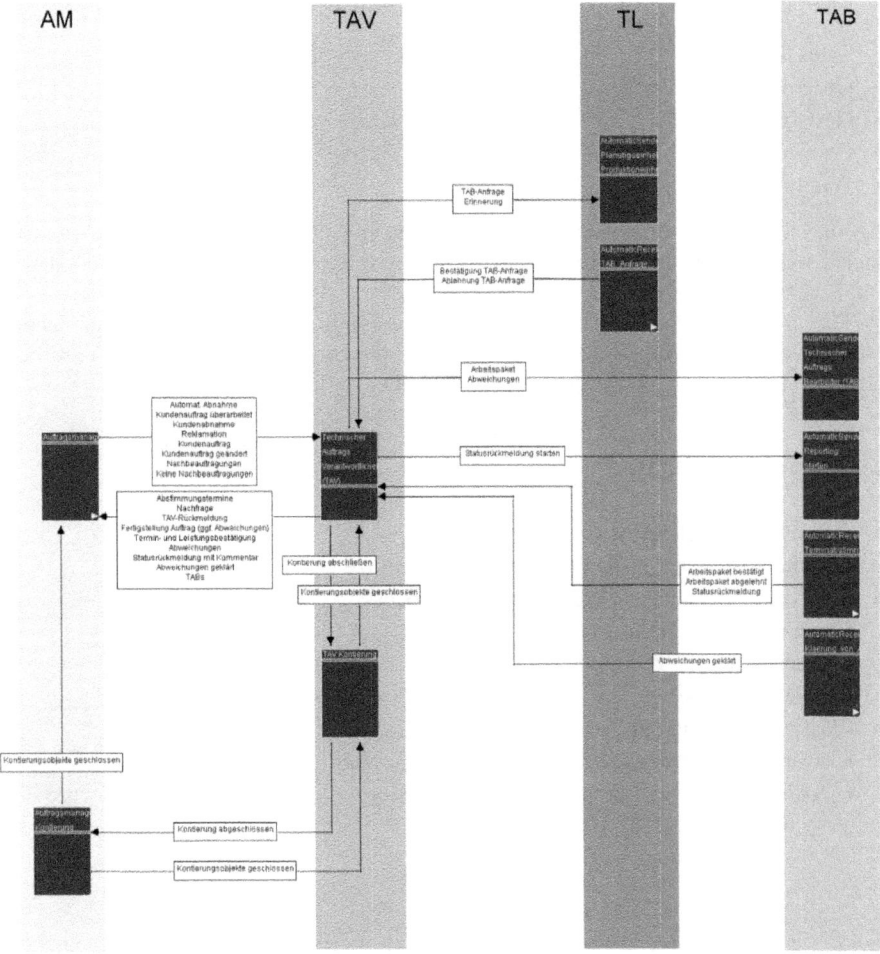

Fig. 2. S-BPM-based order control

names of the organizations which are involved in the order-control process. These boxes represent the embedding of the process into the organization, as actually being implemented.

8 Implementation

The implementation was divided into two phases: a pilot phase and a production phase. In the pilot phase process was thoroughly evaluated. One of the goals of the pilot phase was to become familiar with the usage of the new software, and the newly defined order control process.

The pilot phase started in March 2009 and involved 5 departments:

- Sales department
- Order control department
- Network group
- Firewall group
- Windows planning group

Around 150 improvements were identified during the pilot phase. These improvements were mainly in conjunction with process adaptations, but also included some improvements to the jCOM1 S-BPM Suite.

The process is currently in production, involving almost the entire organizations (while ensuring high-quality work). Training courses (approx.160 persons) are being held, in order to develop the core business further.

10 Lessons Learned

- Plan as much validation time as possible
  - Take advantage of the opportunity to optimize
- Involve the persons associated with the process
  - improvements are possible
  - discussions are required
  - take your time
- Existing tools and applications can be integrated very easily
  - available interfaces are very helpful to integrate existing applications
  - SAP functions can be integrated very easily with the interfaces already provided by the JCOM1 S-BPM Suite.

Part III

Penetration Perspectives

Potential Building Blocks of S-BPM

Hagen Buchwald

Institute AIFB, Karlsruhe Institute of Technology (KIT)
Kaiserstraße 12, 76128 Karlsruhe, Germany
hagen.buchwald@kit.edu

Abstract. This contribution sketches a roadmap on how to proceed in establishing S-BPM from a technological, community and methodological point of view. For each strand a fundamental set of activities considered to be crucial for a sound penetration of development and research is discussed.

Keywords: subject-orientation, Abstract Data Type (ADT), Abstract State Machines (ASM), object-oriented design (OOD), subject.

1 Introduction

The objective of this constitutional conference was to gather a mixed business and scientific community in order to discuss the need for a new paradigm of how to do Business Process Management in the future. The author of this article himself comes from the business side and returned in October 2008 to the KIT to do his PhD and to teach students in the paradigms of object-oriented programming and subject-oriented Business Process Management (S-BPM). Hence, his perspective reflects the application and concept perspective in a balanced way.

In the following the question of how to establish S-BPM by means of so-called building blocks is addressed. These constituents of successful research and development penetration are specified, including a practical definition of S-BPM.

2 How to Establish S-BPM

Before starting work, one should clearly define the desired results he or she wants to achieve. The desired results for establishing S-BPM look like this:

H. Buchwald et al. (Eds.): S-BPM ONE, CCIS 85, pp. 123–135, 2010.

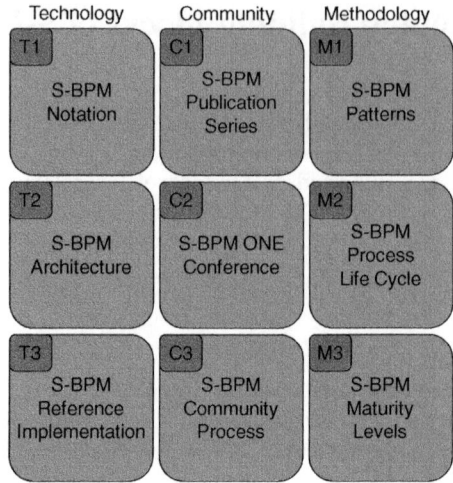

Fig. 1. Potential building blocks of a roadmap to S-BPM

There are three streams of activities:

1. Technology – the T-Stream
2. Community – the C-Stream
3. Methodology – the M-Stream

We can refine each of the streams to building blocks:

1. T-Stream
   - T1 – S-BPM Notation
   - T2 – S-BPM Architecture
   - T3 – S-BPM Reference Implementation
2. C-Stream
   - C1 – S-BPM ONE Publication Series
   - C2 – S-BPM ONE Conference
   - C3 – S-BPM Community Process
3. M-Stream
   - M1 – S-BPM Patterns
   - M2 – S-BPM Process Life Cycle
   - M3 – S-BPM Maturity Levels

A building block represents a set of deliverables which serves a specific purpose in order to establish S-BPM in the sciences and business communities. Working on each building block should allow answering the following four key questions:

1. WHAT is S-BPM?
2. WHY is there a need for S-BPM?

3. HOW does S-BPM work?
4. SO WHAT – what is the implication of S-BPM?

We can also group the building blocks according to the question(s) they might help answering:

1. WHAT?
   - M3 – S-BPM Maturity Levels
   - C1 – S-BPM ONE Publication Series
   - C2 – S-BPM ONE Conference
   - C3 – S-BPM Community Process
2. WHY?
   - C1 – S-BPM ONE Publication Series
   - C2 – S-BPM ONE Conference
3. HOW?
   - C1 – S-BPM ONE Publication Series
   - C2 – S-BPM ONE Conference
   - C3 – S-BPM Community Process
   - T1 – S-BPM Notation
   - T2 – S-BPM Architecture
   - T3 – S-BPM Reference Implementation
   - M1 – S-BPM Patterns
   - M2 – S-BPM Process Life Cycle
4. SO WHAT?
   a. C1 – S-BPM ONE Publication Series
   b. C2 – S-BPM ONE Conference

The following sections will briefly define the specific purpose each building block serves in the process of establishing S-BPM in both, scientific and business communities.

One building block – S-BPM Maturity Levels (M3) – will be presented in more detail, as it gives the answer to the major question: WHAT is S-BPM? The answer to this question is key, namely to get all the terms and definitions we need to answer the question of WHY, HOW and SO WHAT in a coherent way.

3 T1 – S-BPM Notation

This building block gives an answer to the question of how S-BPM models should be visualized for design purposes.

The criteria to review the different alternatives should be:

1. The S-BPM Notation should be as closely aligned as possible with existing standard notations. As an example, S-BPMN (S-BPM Notation) could be a mere subset of BPMN (see www.bpmn.org).
2. The S-BPM Notation should be as simple as possible. As an example, the number of different shapes to be used should be about half a dozen, not more.

3. The S-BPM Notation should be readable for business people. As an example, a business expert working on the process should be able to recognize the designed behavior of the subject describing his or her own role in the process, after a 15 minutes' introduction to the S-BPMN shapes and their meaning.
4. The S-BPM Notation should be usable for business people to modify the behavior of the subject which describes their own role in the process, after a 60 minutes' introduction on how to design the behavior of a subject.

The Parallel Activity Specification Schema (PASS) [1] is a first step in the direction of such an S-BPM Notation, as it fulfills the criteria 2 and 3 quite well, while meeting criterion 4 partially, and violating criterion 1 entirely.

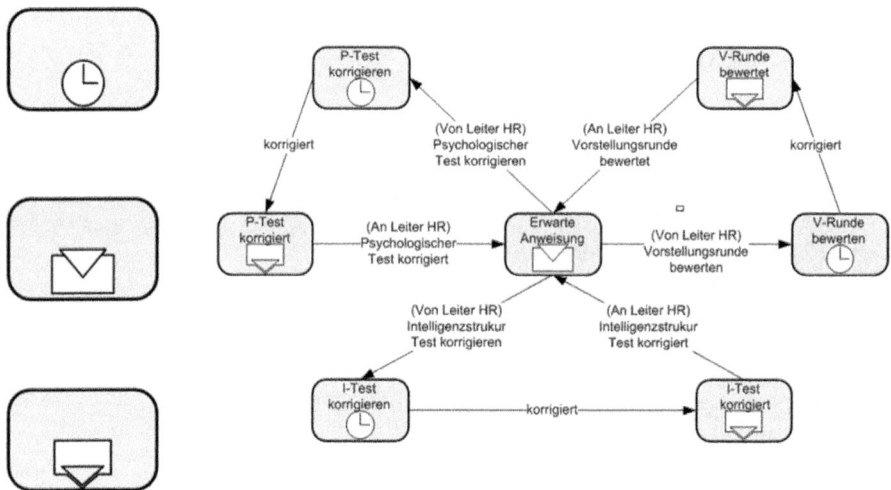

Fig. 2. S-BPM Notation

4   T2 – S-BPM Architecture

This building block gives an answer to the question of how S-BPM systems should be designed to meet the requirements of a full-fledged S-BPM platform.

The criteria to review the different alternatives should be:

1. The S-BPM Architecture should consist of different layers in order to allow S-BPM platform vendors to adhere to the S-BPM standards and at the same time have the freedom to offer a unique value proposition. As an example, S-BPM platform A and B consist of three layers:

   a. Descriptive layer, interpreting the S-BPM models design in S-BMPN.
   b. Logical layer, optimizing the interpreted model by algebraic transformations which depend on the internal representation of the S-BPM model.
   c. Physical layer, executing the optimized internal representation of the S-BPM model.

The logical and the physical layer could be completely different in S-BPM platform A as compared to S-BPM platform B, offering the customer more performance or lower resource consumption.

The JCOM1 BPM Suite (www.jcom1.com) is a first step in the direction of such an S-BPM Architecture, as it separates the descriptive layer from the physical layer, but yet it still lacks the clearly separated logical layer.

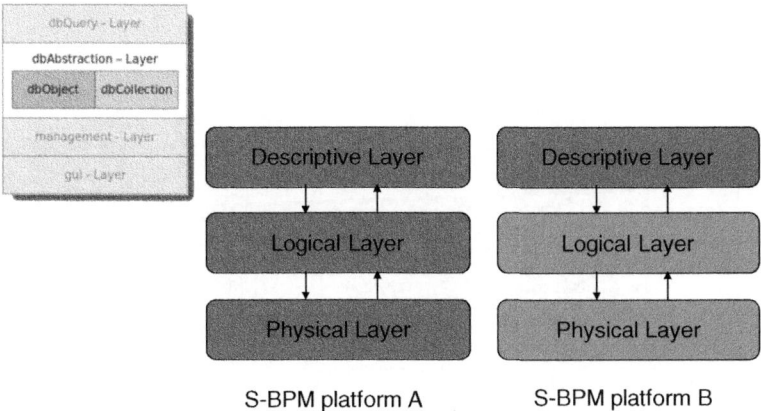

Fig. 3. S-BPM layered architecture

5   T3 – S-BPM Reference Implementation

This building block gives an answer to the question of how S-BPM platforms could be implemented, serving as both, a functional and performance benchmark to S-BPM vendors.

The criteria to review the different alternatives should be:

1. The S-BPM Reference Implementation should adhere to all maturity levels presented in building block M3.
2. The S-BPM Reference Implementation should be a functional benchmark offering all the functionality (services) the S-BPM standard describes.
3. The S-BPM Reference Implementation should scale in a linear way on a multi-core hardware platform proving the inherent parallelism of the S-BPM paradigm.
4. The S-BPM Reference Implementation should be open source.
5. The S-BPM Reference Implementation should be a robust standard for developing S-BPM systems, but not a competitor in the S-BPM platform market for running mission critical business processes.

Currently, there exists no such open source S-BPM Reference Implementation. As a good example, the Java 2 Platform Standard Edition of the late 1990s showed how such a reference implementation could serve as an attractor to attract the different implementations of application servers in the market to the standard.

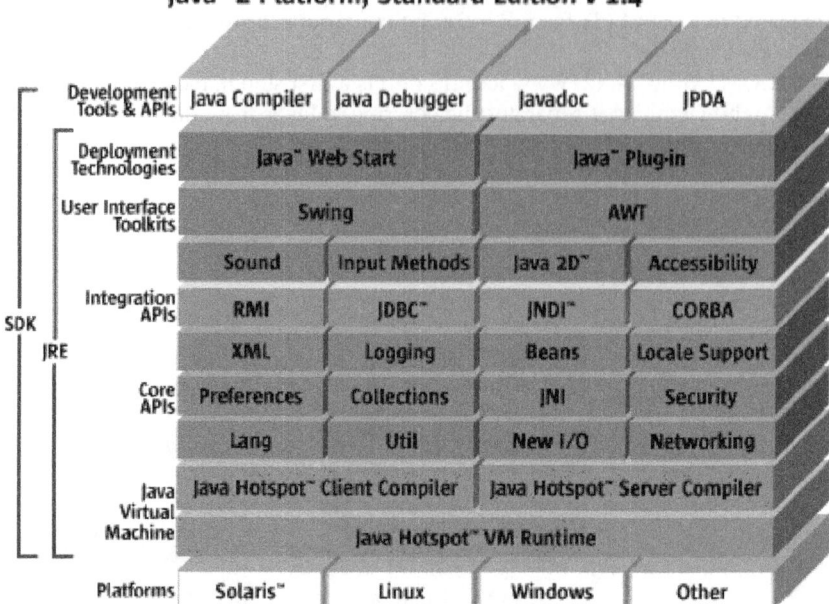

Fig. 4. S-BPM Reference Implementation

6   C1 – S-BPM Publication Series

This building block broadcasts the answers of what S-BPM is, why we need S-BPM, how it works and what S-BPM will change in both, scientific and business communities. The criteria to review the different alternatives should be:

1. The S-BPM Publication Series should be published by a well-known publisher.
2. Searching S-BPM publications at amazon.com, amazon.de and amazon.jp should lead to at least one screen full of respective hits.
3. Searching for a definition of S-BPM at wikipedia.org should lead to results in at least English, German and Japanese, the languages of the markets which are – at the moment – the most important markets for S-BPM.

Currently, there are neither specifications at wikipedia.org which accurately define S-BPM, nor books which could be purchased at amazon.com, amazon.de or amazon.jp.

7   C2 – S-BPM ONE Conference

This building block brings together both, the scientific and business S-BPM communities. The objective is to review and discuss how far we have proceeded on the S-BPM roadmap and which milestones need to be achieved for the next year.

The criteria to review the different alternatives of how to organize this conference should be:

1. The S-BPM ONE Conference should be attractive to both, scientific and business communities. For example, it could be split up into a two-day conference addressing the scientific community on the first day and adding the business community on the second day.
2. The S-BPM ONE Conference should take place at a renowned university or institution in an annual cycle.

The constitutional S-BPM ONE Conference 2009 took place at the Karlsruhe Institute of Technology in Karlsruhe, Germany, the fusion of the University of Karlsruhe, which is well-known for both computer science and business administration in Europe, and the former German National Nuclear Research Center.

8 C3 – S-BPM Community Process

This building block gives an answer to the question of how S-BPM standards should evolve in a transparent and involving way, and which board should subsequently set the standards.

The criteria to review the different alternatives should be:

1. The S-BPM Community Process should adopt best practices from the World Wide Web community. As an example, the Java Community Process (JCP) is such an established and working mechanism for developing standard technical specifications for Java technology.
2. The S-BPM Community Process should be easy to understand, transparent in its execution, and fast in taking decisions.
3. The S-BPM Community Process should take both, the scientific community and business community into account, serving both in the most efficient way.

At the moment, there is no such S-BPM Community Process (S-BPMCP).

9 M1 – S-BPM Patterns

This building block gives an answer to the question of how S-BPM modelers could use design patterns in order to establish best practices for S-BPM models.

The criteria to review the different alternatives should be:

1. The S-BPM Patterns should be split into categories, making it easy to categorize them. As an example, workflow patterns are organized in categories like control flow, resource, data and exception handling.
2. The S-BPM Patterns should be as few as possible. As an example, it was possible to show that the 43 control flow workflow patterns could be mapped to only 25 S-BPM patterns based on an extended version of PASS.

First steps have been taken to research S-BPM patterns on an experimental base, but not on a theoretical base.

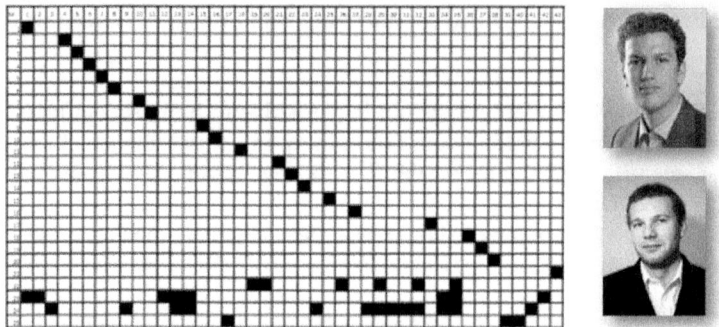

■ Nils Tölle and Norbert Graef showed in their Bachelor-Thesis 2009 [2],
  that it is possible to reduce van der Aalst's (see [3])
  43 workflow patterns to 25 S-BPM patterns.

Fig. 5. S-BPM Patterns

10 M2 – S-BPM Process Lifecycle

This building block gives an answer to the question of how S-BPM projects could be conducted using the advantages of the S-BPM paradigm in an effective and efficient way.

The criteria to review the different alternatives should be:

1. The S-BPM Process Lifecycle should be derived from the core definition of what S-BPM is.
2. The S-BPM Process Lifecycle should be incremental, iterative and value and risk driven.
3. The S-BPM Process Lifecycle should involve the customer, where customer means the employees working in the business process.
4. The S-BPM Process Lifecycle should focus on capturing the as-is process first and improving it by a continuous improvement process based on measured facts and employee observations of disturbing factors and improvement potential in the business process.
5. The S-BPM Process Lifecycle should be a means to embrace complexity instead of an attempt to erase complexity. Complexity in business processes is a consequence of the way individuals work in the process. Individuals' behavior is individual, as it adheres to the "success patterns" which means the habits of the employees working in the business process.

There exist first success stories on how to conduct S-BPM projects, trying to combine best practices of well-known BPM process lifecycles like Six Sigma or KAIZEN with maturity models like CMMI and early warning error detection techniques like Complex Event Processing (CEP).

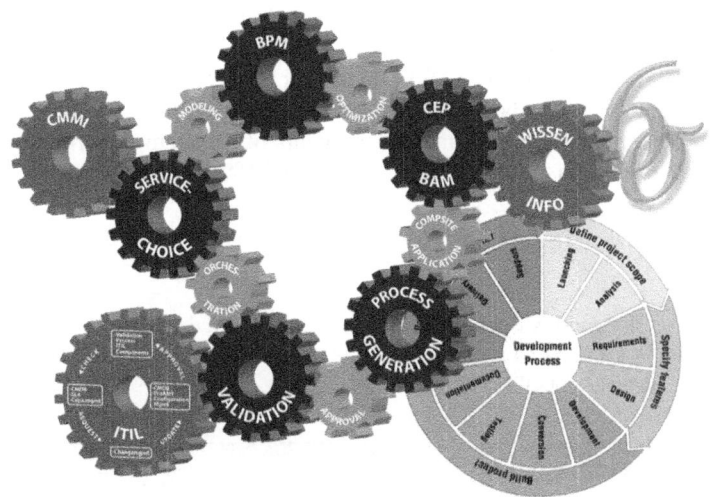

Fig. 6. S-BPM Process Life Cycle

11   M3 – S-BPM Maturity Levels

This building block gives an answer to the question of what S-BPM exactly is and which criteria an S-BPM platform must fulfill.

The criteria to review the different alternatives should be:

1. The S-BPM Maturity Levels should be incremental. Basic requirements should define a low level of maturity, whereas high sophisticated requirements should define a high level of maturity.
2. The S-BPM Maturity Levels should be based on a clear definition of what S-BPM exactly is which means that there should already be an existing theoretical concept as a fundament.

Since we consider it as the crucial building block to provide a solid foundation for S-BPM, we will investigate this building block in more detail.

We could regard S-BPM as the next evolutionary step of programming languages. Starting with the "predicative" (imperative, procedural) paradigm, the object-oriented paradigm evolved from the observation that it is not the functions ("predicates") which are stable in the lifecycle of a software system, but rather the data structures.

Bertrand Meyer descibes in [4] the consequences derived from this observation, as seven maturity levels have to be mastered by an object-oriented programming language in order to be actually object-oriented:

1. Object-based, modular structure: Systems are modularized on the basis of their data structures.
2. Data abstraction: Objects must be described as implementations of abstract data types (ADTs).

3. Garbage Collection: Unreferenced objects should be de-allocated automatically by the underlying runtime system.
4. Classes: Classes are implementations of abstract data types (ADTs).
5. Inheritance: A class can be defined as a reduction or extension of another class.
6. Polymorphism and dynamic binding: Elements of a system may reference objects of more than one class, and routines may have different implementations in different classes.
7. Multiple inheritance: You may declare classes which inherit from more than one parent class.

Consequently, Bertrand Meyer [4] defines object-oriented design (OOD) as "the design of software systems as structured compilation of classes, i.e. implementations of abstract data types (ADTs)".

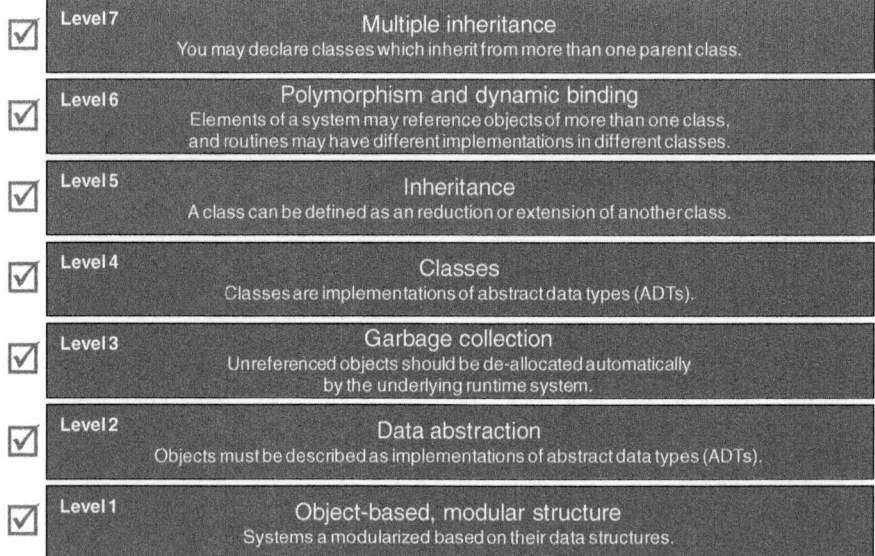

Fig. 7. OOP Maturity Levels

Transferring this understanding to S-BPM, we could state the following: It is not only the data structures which are the stable element in software systems. Especially for software systems implementing business processes, we observe that the habits of the people working in the business process are amazingly stable. As a consequence, even a so-called 'to-be' (i.e. envisioned) process is not executed in one standard way, but in a number of variations, depending on which individuals executed the different parts of the business process. It is a combinatorial explosion we face if we examine how the different instances of the same process were performed. No instance resembles another due to different individuals performing the process. Having a business process divided up into 10 workstations (which is not many) and 5 employees (which again is not many) working at each workstation, you get 100.000 variations of the "same" business process. The only stable thing in this jungle of variations is the set of employee habits!

What does this mean for an S-BPM Maturity Level model?

On the basis of Bertrand Meyer's definition of seven maturity levels, we now try to define seven maturity levels which all have to be mastered by a subject-oriented "programming" language in order to be really and truly subject-oriented. Therefore we replace:

- Object by subject: A subject is not passive like an object, but active like a protagonist in a plot. An object is executed by a business process; a subject executes the business process.
- Classes (of objects) by classes (of subjects): An object is an instantiation of a class of objects, which is like a blueprint specifying how an object can be executed. A subject is an instantiation of a class of subjects, which is like a blueprint specifying how the subject behaves in order to execute a business process. The implementation of this behavior differs from class to class, as this behavior is based on habits.
- The concept of abstract data types (ADTs) by the concept of abstract state machines (ASMs) as described in [5]: An abstract data type (ADT) describes a set of services a specific class of objects exposes to its environment and the contract which must be fulfilled in order to use these services properly. An abstract state machine (ASM) describes the behavior a specific class of subjects exposes to its environment and the contract which must be fulfilled in order to cooperate with these subjects properly.
- Garbage collection by linear scalability: The technical aspect of being able to detect and automatically destruct objects which have become memory garbage was crucial for long-running mission critical object-oriented systems. The technical aspect of being able to scale in a linear way by adding new cores to the hardware will be crucial for mission-critical, on-demand business process execution which suffers from volatile loading induced by a volatile market demand.

This mapping leads us to the following definition of the S-BPM Maturity Levels:

1. Subject-based, modular structure: Systems are modularized on their subjects' behavior (habits).
2. Behavior abstraction: Subjects must be described as implementations of abstract state machines (ASMs).
3. Linear scalability: The more cores the underlying hardware offers, the more subjects can act in parallel in order to respond to higher market demand.
4. Classes: Classes are implementations of abstract state machines (ASMs).
5. Inheritance: A class can be defined as a reduction or extension of another class.
6. Polymorphism and dynamic binding: Elements of a system may reference subjects of more than one class, and behavior may have different implementations in different classes.
7. Multiple inheritance: You may declare classes which inherit from more than one parent class.

Thus, based on Bertrand Meyer's lines of thought which led to the definition of object-oriented design, we can now provide the definition of subject-oriented business

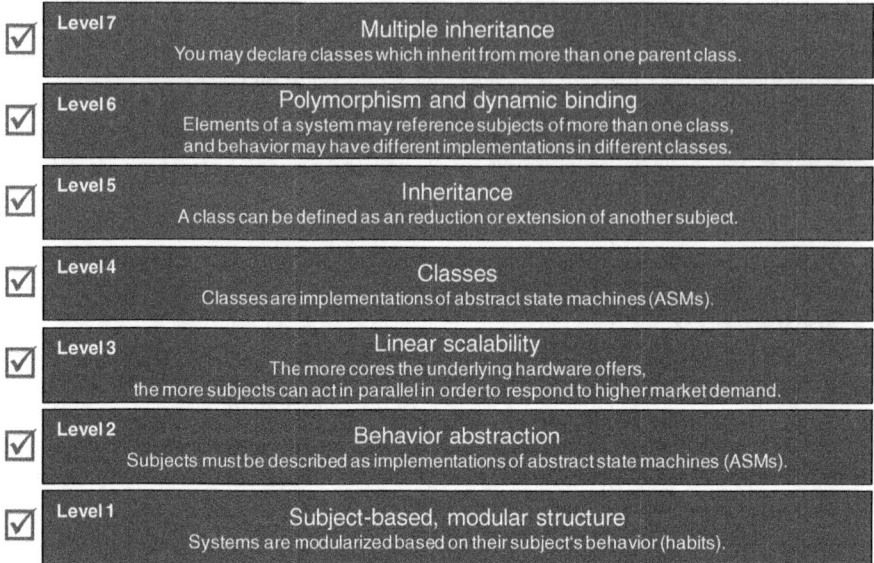

Fig. 8. S-BPM Maturity Levels

process management (S-BPM) as "a design and implementation paradigm for process-based systems as structured compilations of classes of subjects, i.e. implementations of abstract state machines (ASMs)".

12   Conclusive Summary

This article proposes three streams of action consisting of three building blocks of how S-BPM could be established:

1. Technological Stream T
   a. T1 – S-BPM Notation: A simple notation, easy to learn und easy to use, preferably a subset of BPMN.
   b. T2 – S-BPM Architecture: A standard for what an S-BPM platform should contain without specifying how it is done.
   c. T3 – S-BPM Reference Implementation: An open source implementation of an S-BPM platform which proves that there exists at least one working implementation of the S-BPM architecture which could be used as a functional (not performance) benchmark.
2. Community Stream C
   a. C1 – S-BPM Publication Series: A series of books and articles explaining the idea and the value proposition of S-BPM.
   b. C2 – S-BPM ONE Conference: An annual conference gathering the growing worldwide S-BPM community (science and business) for two days.
   c. C3 – S-BPM Community Process: A clearly defined process of how decisions concerning the S-BPM standard are prepared, taken und published.

3. Methodological Stream M

    a. M1 – S-BPM Patterns: Design Patterns which ease the communication of best practices found in S-BPM process models.

    b. M2 – S-BPM Process Lifecycle: An approach on how to proceed in S-BPM projects from requirement capturing up to acceptance tests in an incremental, iterative and value und risk driven way.

    c. M3 – S-BPM Maturity Levels: A consequent list of criteria an S-BPM system must fulfill in order to be allowed to carry the label S-BPM.

The building block M3 – S-BPM Maturity Levels - was introduced in more detail as it offers the crucial answer of what S-BPM is. The proposal of a definition given in this article is: S-BPM is a design and implementation paradigm for process-based systems as structured compilations of classes of subjects, i.e. implementations of abstract state machines (ASMs). Just as objects are instantiated from classes of objects in object-oriented languages, subjects are instantiated from classes of subjects in subject-oriented languages.

Using the metaphor of natural language, S-BPM is a language that allows us to describe as-is business processes in terms of subjects, predicates and objects, i.e. natural sentences.

The key value proposition of S-BPM is the focus on the subject: Subjects in business processes are mainly human beings. S-BPM allows enterprises to transform as-is business process by continuous improvement into standards (best practices) without tackling the variety of habits in the micro-cosmos of a subject. Thus, S-BPM projects create high user acceptance and are more likely to be successful and therefore fulfill their mission: To improve the productivity of business processes which are based on the division of labor.

Acknowledgements. Special thanks go to Dr. Albert Fleischmann for stimulating discussions on the very first ideas of this subject during several meetings starting in 2003, and to Professor Andreas Oberweis and Professor Detlef Seese of AIFB, KIT for their deep questioning in trying to explore the real nature of S-BPM.

References

1. Fleischmann, A.: Distributed Systems. Software Design & Implementation. Springer, Berlin (1994)
2. Graef, N., Tölle, N.: Evaluation, Mapping and qualitative Reduction of Workflow Pattern (Control-Flow) - A practical, subject-oriented procedure by applying the jCOM1 suite (original title in German: Evaluation, Mapping und qualitative Reduktion von Workflow Pattern (Control-Flow) - Eine praxisnahe, subjektorientierte Vorgehensweise durch Verwendung der jCOM1 Suite), Bachelor-Thesis, submitted at May 28th 2009 at the Institute AIFB, Karlsruhe Institute of Technology (KIT) (2009)
3. Russell, N., ter Hofstede, A.H.M., van der Aalst, W.M.P., Mulyar, N.: Workflow Control-Flow Patterns: A Revised View, BPM Center Report-06-022, BPMcenter.org (2006)
4. Meyer, B.: Object-Oriented Software Construction. Hanser, München (1986)
5. Börger, E., Stärk, R.: Abstract State Machines – A Method for High Level System Design and Analysis. Springer, Berlin (2003)

Quo Vadis, S-BPM? The First World-Café on S-BPM Developments

Christian Stary

JKU, Business Informatics – Communications Engineering
Freistädterstraße 315, 4040 Linz, Austria
Christian.Stary@JKU.AT

Abstract. World cafés allow for collective reflection and brain writing. S-BPM is still in its beginnings with respect to implementing the concept in organization designs and focusing on communication rather than on functions and the corresponding type of flow control. A World Café has been set up for reflecting current achievements and further developing S-BPM. 4 thematic areas have been selected from the topics addressed by the participants throughout the workshop, and discussed in 4 round table sessions. Some café discussions have led to further scoping S-BPM and agreed conclusions while others have triggered intense debates on dedicated S-BPM aspects, e.g., how to educate S-BPM users for empowerment. This report summarizes the process and findings of the first World Café on S-BPM development.

Keywords: World Café, collective intelligence, subject orientation, standardization, education, business model creation, organizational culture.

1 Introduction

Due to its nature S-BPM [2] is an approach that might re-invent business process modeling and management. In any case, shifting the focus of BPM towards communication and information exchange requires substantial re-thinking and significant effort in re-engineering business. Being aware of this fact the organizers of S-BPM'09 did not only get together researchers, practitioners, and concerned developers, they also trusted that when these informed presenters and discussants would be given the opportunity to voice their motivation for aspiring to be S-BPM proponents, their curiosity for improving S-BPM as a technique, and establishing it as organizational paradigm amplifies. By having S-BPM interested persons connect with their own experiences and leadership potentials, the World Café (www.theworldcafe.com) should inspire to find deeper meaning in modeling and organizational development practices, if not business pursuits.

The World Café was set up at the end of the day of the workshop, as it should help to wrap up what has been said and what could direct further S-BPM developments. Since it should be based on the most prominent questions of the workshop, we needed to collect the various thoughts that came up in the course of discussing the contributions. Here is a list of questions that have been raised:

H. Buchwald et al. (Eds.): S-BPM ONE, CCIS 85, pp. 136–147, 2010.

- How to implement S-BPM as a paradigm of new/traditional (since human) thinking (in natural language)?
- How to project transitions from traditional BPM [5] to S-BPM?
- What are the most pressing features of S-BPM?
- S-BPM – a semantic technology?
- How to teach (knowledge) workers S-BPM?
- What is a successful S-BPM business case?
- How does the Internet of services affect S-BPM?
- What could an S-BPM standard look like?
- How to implement culture-sensitive S-BPM in a few days?

From these questions it becomes evident that user education and qualified user involvement in organizational development is considered central to effective S-BPM. Once S-BPM is established for organizational development a standard notation or language for S-BPM might become a crucial issue. It might trigger the further use of S-BPM, in particular in cross-organizational business development and networking projects. However, cultural (diversity) issues, e.g., migrating Indian and European organizations, might need to be addressed explicitly for successful business mergers. Accordingly the following four topics have been selected for the first S-BPM World Café:

1. How to educate for user empowerment?
2. What about standardizing S-BPM?
3. How to establish a/the S-BPM business?
4. Organizational culture and S-BPM

Each of these topics has been assigned to a particular table of the World Café in the commons of the workshop venue. Subsequently, participants interested in elaborating one of the above listed issues volunteered for hosting respective tables. After 4 rounds of discussions the hosts shared the intelligence collected at their table with the others.

In the following sections we briefly introduce the World Café as a technique for collecting and sharing (S-BPM) intelligence through dialog and conversation, before we detail the results from the round table discussions. The conclusive summary wraps up the findings and complements them by related scientific or expert findings, in order to position the achievements according to the state of the art in (S-)BPM.

2 S-BPM World Café One: Conversation Matters

In this section we introduce the World Café as an effective means for collecting intelligence in participatory knowledge management settings. The World Café (www.theworldcafe.com) is an effort towards stimulating creative conversation about questions and issues that matter in communities. Using the Café as a *methodology* and as a *metaphor* offers a practical and innovative way to cultivate both the knowledge required to thrive as S-BPM scholars and the experience needed to trigger future developments in respective application fields. As such, it provides means of orientation and development for the S-BPM community.

As knowledge sharing is a key element of knowledge management and for organizational success, the World Café should help to get S-BPM practitioners, researchers, and developers to talk openly to one another other about their specific (corporate) interests, opportunities and responsibilities. It is an effective vehicle for opening up conversations and discussions that lead to sharing expertise. One way of thinking of the World Café is as a tool that helps to share so-called tacit knowledge, i.e. knowledge that has not been documented and codified so far. When used within teams or Communities of Practice to question entrenched assumptions, it helps facilitate learning from others and gain a deeper collective understanding of a subject through conversation.

World Café interaction is not just about talking and networking though these are secondary benefits but allowing people to engage each other in dialog on S-BPM with the aim of learning from each other rather than entering into unproductive (since mostly dogmatic) debate and attempting to impose their views (or dogmas) on the other which invariably end in destructive conversations and frustration of participants. Specifically a Knowledge Café can:

- help to gain improved understanding of a complex issue, such as mapping perceived organizational realities to models
- get buy-ins for a new initiative, such as for S-BPM-based organization development
- flush out problems and issues in an initiative or project especially ones of lack of communication that can then be acted on and resolved, e.g., when educating organization designer in S-BPM
- help build consensus around a proposed plan of action, such as the future activities to establish S-BPM as a paradigm
- improve the way that people work together by gaining a deeper understanding of each others perspectives on issues, e.g., by bringing together practitioners and developers
- allow to jointly develop a policy document, e.g., when intending to share material on S-BPM education
- serve as a tacit transfer mechanism between young and senior researchers, practitioners or developers, e.g., observing senior consultant when designing S-BPM projects
- facilitate more widely sharing individual expertise, in particular through collecting group intelligence on specific issues, such as S-BPM notation design
- help merge two organizational cultures or work practices, e.g., object-oriented business process modeling techniques and S-BPM
- improve inter-personal relationship and thus ability to work together effectively, e.g., forming liaisons to design and implement a research agenda on S-BPM

In effect the Knowledge Café is an easy, low cost way to make knowledge sharing happening. The café is built on the assumption that people already have within them the knowledge and creativity to confront even difficult challenges. Given the appropriate context and focus, it is possible to access and use knowledge about what is important in a field.

As such, the Knowledge Café is a *methodology* for creating a living network of collaborative dialog around critical issues and questions that matter for a concerned community (as in our case the S-BPM proponents). It is a *metaphor* that enables members of a community (also by inviting new people) to gain new insights to make a difference in the participants' professional (and in many cases also personal) thinking and acting.

The Café format is flexible and adapts to different circumstances. We used the following guidelines in combination to foster collaborative dialog, active engagement and constructive possibilities for S-BPM action.

- *Clarify the Purpose* – Paying attention early to the reason bringing people together helps with the facilitation of questions and highlighting the parameters that are important to achieve a certain purpose. We have transcribed the lively discussions to overview the set of issues to be tackled collectively.
- *Create a Hospitable Space* – When people feel most comfortable to be themselves, they do their most creative thinking, speaking and listening. The facilitator aimed to create a space that feels safe and inviting. The workshop venue provided an ambience for constructive social interaction.
- *Explore Questions That Matter* – Café conversations are as much about discovering and exploring powerful questions as they are about finding effective solutions. A Café may explore a single question only, or several questions may be developed to support a logical progression of discovery throughout several rounds of dialog. The selection of questions was based on the criteria of self-containment and base line specifity.
- *Encourage Everyone's Contribution* – It is important to encourage everyone in the meeting to contribute their ideas and perspectives, while also allowing anyone who wants to participate by simply listening to do so. Participants have been asked to assign themselves to their topic of interest for the first round table discussions. It helped twofold: The initial round for each of the topics brought up the most urgent themes and lines of discussions, whereas the subsequent rounds revealed different perspectives on the already documented content on the table.
- *Connect Diverse Perspectives* – The opportunity to move between tables and meet new people contributes to one's thinking and links individual discoveries to ever-widening circles of thought. The hosts did not force participants to link their thoughts immediately to the existing table content. However, the participants could develop a picture of what has being said so far.
- *Listen Together and Notice Patterns* – Participants are invited to listen with an openness to be influenced by the speaker and for deeper questions, patterns and insights. They are also asked to listen for what is not being spoken about The hosts were invited to engage all table guests in providing input and elaborate their thoughts (in the context of the already documented content on the table). It facilitated the identification of patterns.

According to the guidelines for Café conversations 4 or 5 people are located at small Café-style tables or conversation circles. Then progressive rounds of conversation of approximately 20-30 minutes each are set up. Upon completing the initial round of conversation at the latest, 1 person is asked to remain at the table as the "host" while the others serve as travellers or "ambassadors of meaning." The travellers carry key ideas, themes and questions into new conversations. When searching for a new table, travellers can be challenged to co-locate with others who appear 'most different' from them.

The table hosts and Café members are encouraged to write, doodle and draw key ideas on paper at tables. The table host welcomes new guests and briefly shares the main ideas, questions and themes of the prior conversation. The guests are encouraged to link and connect ideas coming from their previous table conversations by listening carefully and building on each other's contributions.

After several rounds of conversation, a period of sharing discoveries and insights in a whole group conversation is initiated. It is in these town hall meeting-style conversations that patterns can be identified, collective knowledge grows and possibilities for action emerge.

3 S-BPM Education = Empowerment of Users

Addressing the question "How to educate for empowering users?" participants of this table have focused on actual user and organization empowerment through S-BPM education. The first topic that has been addressed by the participants was empowerment. Besides questioning the object of empowerment – 'Who should be empowered?' - They experienced a constructive round robin discussion leading to the insight that empowerment should proceed bottom up, from individual S-BPM users to the organizational level. However, it requires both, a high degree of self-organization, and individual confidence ideas are valued at the organizational level, e.g., by responsively propagating constructive inputs for change and innovation.

Empowerment also requires highly motivated S-BPM users. To that respect the participants discussed intensively advantage and differences of various business process modeling approaches and languages. They tackled the issue of how to give proof of S-BPM's effectiveness and efficiency. One proof could be the cost advantage. Successful S-BPM projects are cost-effective, as modeling is a straightforward task. It does neither consume sophisticated preparation or training – every person is trained in natural language communication which lays ground for subject-oriented modeling – nor expensive IT equipment.

Mentioning IT support for modeling brought up the tool aspect. The participants identified flow games to bring processes to life, as they can be experienced individually, and thus, linked to emotions. Such business (process) simulators should be built similar to flight simulators, as their use improves motivation and self-organization. Typical benefits from deepening self-organization at the workplace are handling missing links in business processes or unexpected interrupts. They could be simulated in business process games.

Finally, empowering users requires comprehensive understanding of roles and process didactics. Education involves many different roles. The spectrum spans from researchers bringing up new concepts to users acting along specified processes. Effectively, education requires (subject-oriented) process specifications for each acting role. Moreover, a variety of tools could support educational S-BPM processes. Besides tutorials, style guides, manuals, hypermedia should be used for effective and positive knowledge transfer. It would not only allow different presentation formats

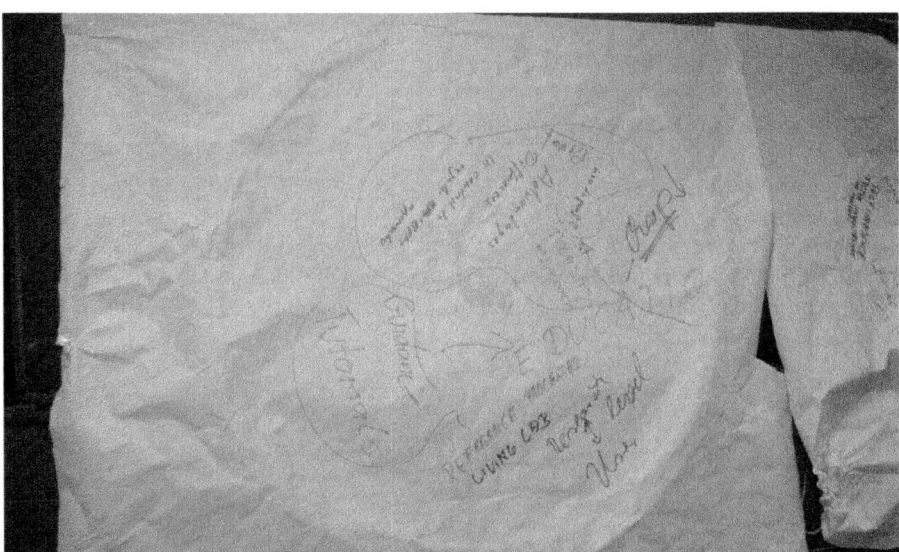

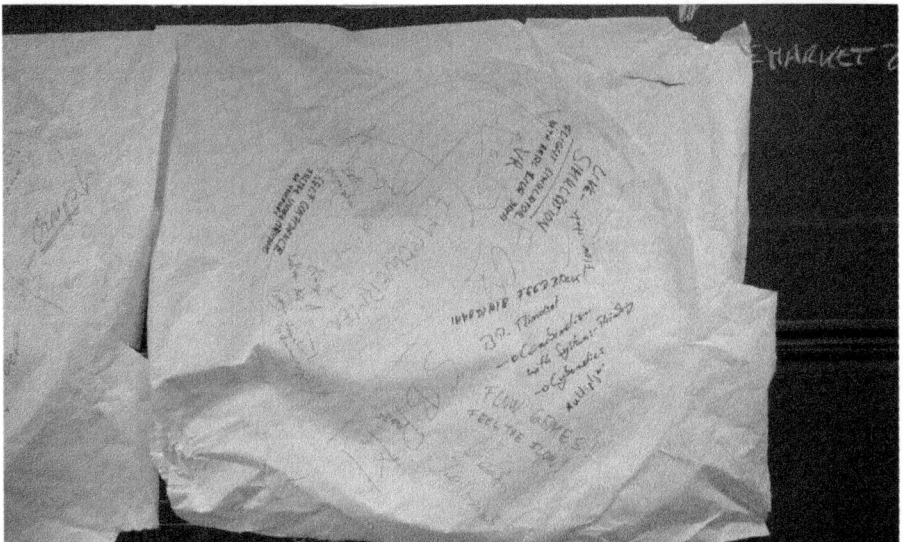

Fig. 1. Collected S-BPM education issues

but also enforce explicit switching between roles, meaning to learn to think in different perspectives.

4   S-BPM Standardization: What and How?

At that table the participants tackled the question 'What could an S-BPM-standard look like?' In that context they brought up some crucial and challenging aspects of S-BPM standardization, including essential questions, such as:

- Why do we need a standard?
- For whom should it be meant?
- Are tool vendors interested in standardization? Which of them?
- Are the customers interested in standardization?

The motivation to create an S-BPM standard could be driven by which BPM standards are de facto available. Is it BPMN, providing a vast amount of diagrammatic symbols and a complicated grammar? Could S-BPM be some kind of 'BPMN light'? If yes, which symbols and rules do we actually need? Such issues lead to the target group(s) of standardization. They might differ according to the (consecutive?) layers of standardization:

1. modeling
2. execution
3. orchestration
4. collaboration

Besides meeting the needs of target users a standard need to have a clear scope, in order to determine its applicability and context of use.

Fig. 2. Collected S-BPM standardization issues

In case tool vendors are interested in S-BPM standardization, it needs to be clarified which standardization organization could support the process and its outcomes (OMG, OASIS, W3C, or others?). The participants discussed several potential strategic partners, such as IBM or Apple that could help to speed up market penetration and product diversification.

5 Business Models for S-BPM-Based Organization Development

Addressing the question "What kind of S-BPM business models could be on the market in 2010?" can be linked to the issue of strategic partnership tackled by the participants discussing standardization. Such collaborations could influence the business model, when S-BPM is considered as a collaborative or networked business endeavor. However, the participants concluded to that respect that a working business model does not require an S-BPM standard.

However, an S-BPM business model should contain all ingredients to generate business with a standardized process management platform. Typical ingredients available that can be expected in 2010 are:

- B2B-processes connecting companies based on a S-BPM suite
- S-BPM application development with reduced costs and reduced risk for customers, due to standardized procedures of the development process
- rearrangement and reuse of IT infrastructure and processes

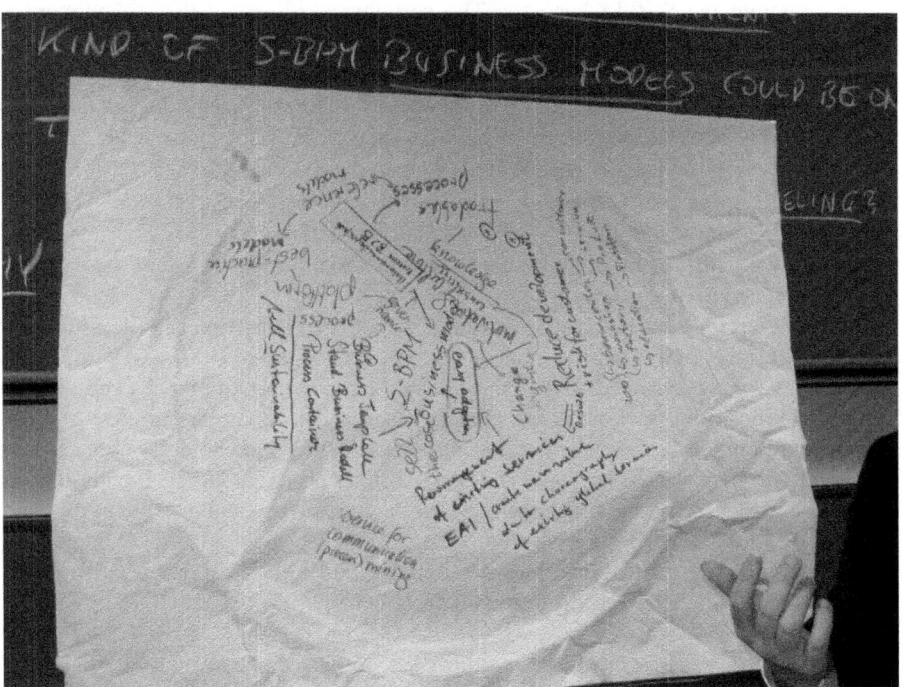

Fig. 3. Collected S-BPM business model elements

The latter comprises connectivity of services, e.g. EAI business objects/services, and rearranging existing IT systems to provide new functionality without re-engineering the systems themselves or building/buying new systems. Reuse in that sense protects investments. Learning from rearrangement projects allows offering process containers and business templates to reuse connectivity. In this way adaptation could be facilitated, providing off-the-shelf working practice or customization. S-BPM introduction should take into account existing infrastructures. For instance, services could connect ARIS [10] to S-BPM business objects. As a possibility to rearrange we can offer process containers.

For implementing such an S-BPM business model the business infrastructure has to be highly developed and easy to access and use. Brain centers should support consultancy in work shop design and factory education, and provide high-level services, such as process mining. Educational support for platform handling and thinking in subjects needs to be established. Finally, the transfer of (process) knowledge should support cultural changes in terms of effective change management.

6 Embodying Organizational Culture in S-BPM

This issue has been intended to determine the extent to which elements of organizational culture is already or can become part of subject-oriented representations or S-BPM projects. The topic can also be easily linked to other table discussions, such as the one on S-BPM business model generation, as organizational culture comes into being play when operating the S-BPM business. The participants discussed the embodiment of organizational culture in S-BPM along the strands of influencing factors, communication, properties of models and the creation of business value through models.

The influencing factors depend on the scope of organizational culture, in particular on the internal and external perspective. The internal view focuses on the organization, management or leadership culture, whereas the external view on customer and partner relationship management. Communication seems to be important when embedding organizational culture in S-BPM. The interaction about processes, via processes, and with customers, in particular their reflection in models, establishes a specification organizational culture.

A follow-up topic of intense conversation was the interdependence of culture and models. Do stakeholders need to change culture once aspects of processes or process models have to be changed, either due to market reasons or internal events? Both might influence how work tasks are accomplished and technologies are used. A culture which is based on strong communication ties among stakeholders, with customers and business partners, is typically reflected in its process models. Exceptions are national barriers for process management, as they stem from culture, business structures, and history (BPM market, law etc.).

The culture of organizations does not only influence the business process model design, but also the behavior, dealing with process models:

(i)   What comes first: Process or culture? Which one can be influenced? In case of reflected cultural factors, the culture might be at stake before processes are at stake.

(ii)   Where in the process is business value created? Depending on the predominant culture of an organization, either value-based or administration-based, business process management might be completely different.

(iii) Even in reflected fits of processes and culture, organizations have to deal with the fact that processes are more dynamic than established or solid culture. Hence, culture and processes need to be re-aligned from time to time.

(iv)   There might be substantial differences in the lifecycle of business process models, depending on how dynamically a company is changing.

A variety of topics with respect to culture and its embodiment into S-BPM has been mentioned, but could not be discussed. One of them was the influence of culture on governance. It might help to explain many S-BPM experiences in terms of responsibility and/or accountability for processes. Will we need a chief culture officer in 2050 or even earlier to mentor S-BPM developments or projects to that respect?

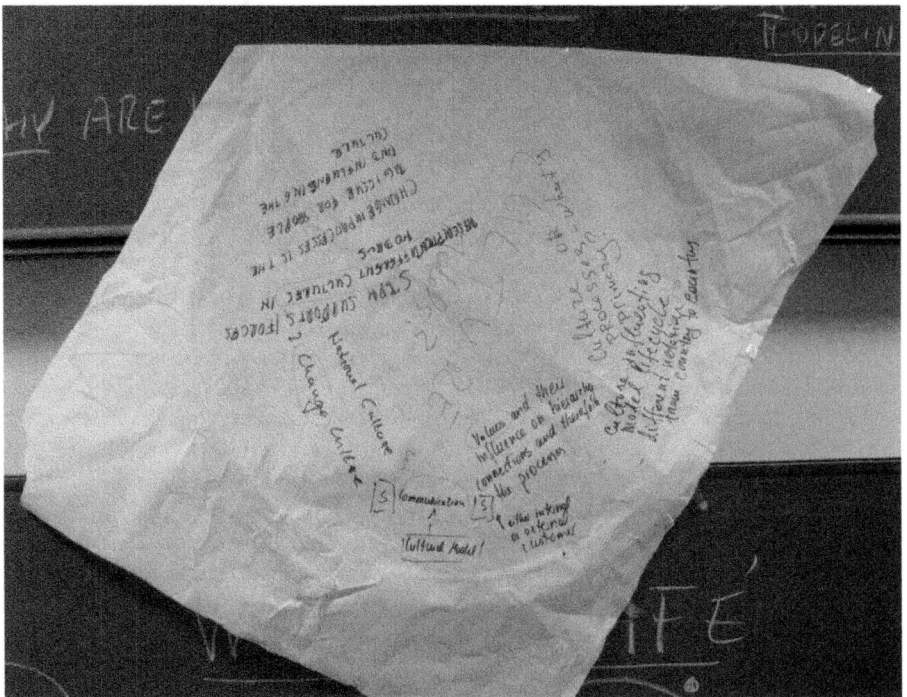

Fig. 4. S-BPM cultural embodiment from an organization's perspective

7   Conclusive Summary

The discussions during the day of the First International Workshop on S-BPM indicated some core issues that should be reflected in a discursive plenary format. Setting up a World Café for that purpose and inviting all participants to join the first conversation of this type justified this impression. It turns out that the technique and technical support are of equal importance to their use. Understanding users and their communication (needs) in an organizational context seems to be crucial for applying and further developing S-BPM.

User education and qualified user involvement in organizational development is considered central to effective S-BPM. It is the perception of stakeholders that accounts for change [6] - interestingly, it is also one of the major business triggers (see also the discussion on generating an S-BPM business model below). In a bottom-up process S-BPM could significantly increase the organizational flexibility which should lead to the required velocity of organizations [11].

Once S-BPM is established for organizational development a standard notation or language for S-BPM might become essential. Standards help to implement concepts of compatibility and reuse [8]. As such, it might increase the use of S-BPM, in particular for cross-organizational business integration in terms of process improvement – only at a first glance: "BPM isn't really all about process improvement, why do organizations purchase it? A review of actual, practical usage of BPM suites today suggests an interesting reality. BPM suites are used primarily for two purposes: application integration and application development. Companies choose to go with BPM when it is the most cost-effective option in one of these areas. Process improvement and optimization is at best considered an ancillary benefit. Notably, Forrester breaks up the market in this way, featuring two different Wave evaluations for *"integration-centric"* BPM and for *"human-centric"* BPM. The latter category represents the application development use case, which typically requires a higher-level support for human-process interaction. This segmentation of the market is becoming increasingly artificial as the two different foci of BPM vendors continue to converge in integrated suites. Nevertheless, Forrester's bifurcation implicitly recognizes that these two use cases are really what BPM is all about." [9]

Consequently, a successful business model has to capture both dimensions. This finding has been seconded by the S-BPM World Café participants. From the integration-centric perspective, the functional application development is out of question. For the human-centric business side, the simulation of integrated or novel processes in an emotionally touching way seems to be crucial. It has to be part of an educational infrastructure that is intertwined with the S-BPM suite. From the application perspective organizational developers and stakeholders (users) need to know the scope of the modeling language (in particular, if an S-BPM standard emerges). It needs to be clarified, what are the essentials of subject-oriented business process modeling ([4], and the current discussions on BPMN at www.bpm-research.com) A proper set of notational elements and a corresponding S-BPM grammar should allow fundamental changes in terms of enterprise transformation to create business value through S-BPM [7].

The participants clearly identified tight connections between business models given by different cultures of organizations and the societal systems organizations are part of. However, for networking and cross-organizational integration the issue of (cultural) diversity has to be tackled explicitly, either through representation in business process models, or through culture-sensitive patterns of behavior, i.e. when dealing with process models. Business process modeling might facilitate substantial organizational changes when capturing the culture [1], such as moving from supply chains to supply networks (cf. www.sudden.biz for the automotive industry). S-SPM clearly addresses one, if not the most crucial aspect of organizational culture:

communication and interaction (see also [2]). In this way, S-BPM could lay ground for a communication-driven sense-and-responsive organizations, as Haeckel [3] termed human *and* technology adaptive enterprises.

References

1. Chakraborthy, D.: Extending the reach of business processes. IEEE Computer 37(4), 78–80 (2004)
2. Fleischmann, A.: What is S-BPM? In: Buchwald, H., et al. (eds.) S-BPM ONE. CCIS, vol. 85, pp. 85–107. Springer, Heidelberg (2010)
3. Heackel, S.S.: Adaptive enterprise: Creating and leading sense-AND-respond organizations. Harvard Business School Press, Cambridge (1999)
4. Havey, M.: Essential Business Process Modeling. O'Reilly, Beijing (2005)
5. Laudon, K.-C., Laudon, J.P.: Essentials of management information systems: Managing the digital firm, 6th edn. Pearson, Upper Saddle River (2005)
6. Lewis, M., Young, B., Mathiassen, L., Rai, A., Welke, R.: Business process innovation based on stakeholder perceptions. Information Knowledge Systems Management 6, 7–17 (2007)
7. Rouse, W.B. (ed.): Enterprise transformation: Understanding and enabling fundamental change. Wiley, Hoboken (2006)
8. S-Cube Consortium: Survey on Business Process Management (2008), http://www.s-cube-network.eu
9. Spurway, K.: The State of BPM: Perspective of an Industry Insider, http://www.bpm.com (10.2.2010)
10. Scheer, A.-W.: ARIS - Modellierungsmethoden, Metamodelle, Anwendungen, 4th edn. Springer, Berlin (2001)
11. Stephenson, S.V., Sage, A.: Architecting for enterprise resource planning. Information Knowledge Systems Management 6, 81–121 (2007)

Author Index